FORAGERS AND FARMERS

Prehistoric Archeology and Ecology
A Series Edited by Karl W. Butzer and Leslie G. Freeman

FORAGERS AND FARMERS

Population Interaction and Agricultural Expansion in Prehistoric Europe

Susan Alling Gregg

The University of Chicago Press
Chicago and London

Susan Alling Gregg is an assistant professor
in the Department of Anthropology at the
University of Washington in Seattle.

The University of Chicago Press, Chicago 60637
The University of Chicago Press, Ltd., London

97 96 95 94 93 92 91 90 89 88 5 4 3 2 1

Library of Congress Cataloging-in-Publication Data

Gregg, Susan A.
Foragers and farmers: population interaction and agricultural expansion in prehistoric Europe / Susan Alling Gregg.
p. cm.--(Prehistoric archeology and ecology)
Bibliography: p.
Includes index.
ISBN 0-226-30735-2. ISBN 0-226-30736-0 (pbk.)
1. Neolithic period--Europe. 2. Agriculture, Prehistoric--Europe. 3. Mesolithic period--Europe. 4. Man, Prehistoric--Food. 5. Europe--Antiquities. I. Title. II. Series.
GN776.2.A1G73 1988
936--cc19

88-21926
CIP

For my parents

Mary Alling Gregg
Samuel R. Gregg

Contents

Figures

Tables

Series Editors' Foreword

Agricultural origins and dispersals pose two separate problems in Old World prehistory that are of great theoretical interest. The first of these has been a subject of close attention for some thirty years, and current opinion favors a long incremental process—millennia of manipulation of and experimentation with potential cultigens and animal domesticates, followed by occasional, then seasonal, and ultimately full-time incorporation of farming traits into what had begun as a broad spectrum of collecting and hunting wild foods. From its earliest steps, to the appearance of "primary" village farming communities as originally postulated by R. J. Braidwood, this transition may have taken five thousand years or more, depending on the location and regional resource availability within the Near East.

The second problem centers on the components and processes of agricultural dispersal beyond the Near East, into Asia, Africa, and Europe. The empirical base for such study is best developed in Europe, which is therefore most suited to develop a sophisticated model as to how farming spreads into new environments already peopled by hunter-gatherers. Since the early writings of V. G. Childe in the 1920s, the traditional explanation has been unconsciously flavored by the European experience in North America. It was argued that farmers, by their superior numbers and technology, progressively advanced and overwhelmed thinly settled, indigenous hunter-gatherers, eliminating or absorbing them, or expelling them to marginal areas. More recently, this model has been reformulated by A. Ammerman and L. L. Cavalli-Sforza as a progressive "wave" of migration sweeping across Europe. This is, of course, an ethnocentric perspective, reflecting the limited historic capacity of northwest Europeans and their New World counterparts to deal with "alien" peoples in a positive way. The Iberian experience in Latin America was much more complex, and in many areas it favored acculturation and assimilation of indigenous peoples, with minimal biological replacement but with a wide variety of interchanges of alternative adaptive traits. Over a span of four centuries it created a cultural mosaic, now blending into more homogeneous societies that are new, rather than European transplants to the Americas.

The antithesis to the wave-of-migration theory has been developed since the 1960s by the "Cambridge school," under the initial stimulus of E. S. Higgs. Its premise was that many of the earliest animal domesticates were not limited to the Near East but were also found in other parts of Eurasia. The transition to farming in southern Europe could therefore have proceeded independently of the Near East. This approach has been refined by G. Barker, who argues for indigenous innovation in combination with later immigration of acculturated farmers—a judicious combination that does not beg the question of why agriculture did, after all, advance from east to west and from southeast to northeast. Nonetheless, this interpretation provides little assistance in explaining the sudden and early appearance of the Neolithic on islands such as Cyprus or Crete, or the total economic discontinuity between the final Paleolithic and Neolithic on the Peloponnissos. It can only ignore the evidence of S. Bökönyi that the earliest Neolithic livestock of the Hungarian Plain represented imported breeds, while later Neolithic animals were derived from the local genetic pool of cattle and pigs, with sheep replaced by better adapted pigs in this marshy terrain.

We are then confronted with two competing general models, one favoring migration (demic diffusion), the other advocating independent innovation or stimulus diffusion, in conjunction with cultural transformation and follow-up migrations. The Iberian experience in Latin America—although at a very different level of social organization—suggests that the answer may be an intricate combination of the two, depending on time and place. This issue transcends European prehistory in that it suggests critical questions for archeologists to ask on other continents. And it goes beyond prehistory as such in that it is fundamental to understanding cultural transformation, in general, and how a repertoire of adaptive behavior is developed, in particular.

Susan Alling Gregg has focused her attention on Central Europe, particularly southwestern Germany, where village farming communities were established by 4500 b.c., and the next millennium or so saw a transition from a bicultural mosaic of Mesolithic hunter-gatherers and Neolithic farmers to a single socioeconomic strand in which a broad range of adaptive experience was combined and transformed.

She argues for protracted, mutually profitable contacts between foragers and farmers rather than expulsion or avoidance of the indigenous folk. In the process, subtle ecological readjustments adapted the Mediterranean-style agricultural system to a comparatively cool-temperate and wet environment, with repeated innovations not anticipated in the coarse-grained and deductive existing models for the "Neolithization" of Europe. Gregg builds from optimal diet models for hunter-gatherers by M. Jochim or B. Winterhalder and E. A. Smith, but she substantially broadens that framework to include farming communities as well as cultural ecological concerns such as population interactions and the complementarity between cultivated and wild resources. She offers an incisive and informed analysis of early agricultural subsistence and an illuminating discussion of diet and dietary needs.

In Gregg's robust and elegant simulation, hunter-gatherers and farmer-herders would have benefited nutritionally in the course of cooperation and competition. She suggests that large agricultural harvests could have been produced regularly. If and when poor spring weather delayed planting, farmers would have needed an additional labor force to plant their crops before the growing season was too far advanced. Local foragers probably provided a pool of emergency labor, and in exchange they would have received wheat from the farmers. A reconstruction of the foragers' seasonal schedule indicates that such cooperation could have been accomplished with few changes in the annual round of subsistence activities. Moreover, the addition of wheat to the forager diet would have reduced the need for fish—a critical, limiting resource factor in determining territory size. Cooperation may have therefore led to a reduction in the territorial requirements for foraging. Because of the periodicity of grain surpluses and the sporadic need for emergency labor, Gregg argues that goods and services were highly elastic commodities. Group interrelationships would presumably have been maintained by incorporating inelastic goods into the social organization and rituals of each.

What emerges is a powerful analytical methodolgy with which early Neolithic archeology can and should be investigated. The author has spent many years working with empirical data in Europe, learning about the local problems of site investigation firsthand, achieving a professional command of the issues and practical problems of research in her region and, above all, acquiring the necessary paleobotanical expertise to identify her own plant materials. The result is a sharp problem-focus and skillful theoretical interpretation that would not be possible without her thorough competence in the subject matter. This well-argued case study is tailored to the specific archeological resolution available in southwestern Germany, but could equally well serve as a framework in which to investigate and interpret the transition to early agriculture in any area where farmers depended on a selection of Near Eastern cultivars and animal domesticates. Gregg makes a particularly effective case that when foragers and farmers occupy a region at the same time, one needs to study both in order to make sense of either.

It is gratifying to have a new volume in the Prehistoric Archeology and Ecology series that deals with Europe, particularly so at a time when a wealth of Neolithic excavation data is being assembled in Germany that has so far found little synthetic interpretation and remains little appreciated in the anglophone literature. But we feel that Susan Gregg's ultimate contribution will be toward enhancing the sophistication of North American archeology students about agriculture as a systemic lifeway, and about the dynamic interactions between foragers and farmers. This is a universal problem in prehistoric interpretation, one with which archeologists excavating in the New World must also cope.

Karl W. Butzer
Leslie G. Freeman

Preface

This book is a pot boiler. It presents a theoretical construct and a methodological approach for exploring relationships that may have existed between foragers and farmers as farming systems became established in Central Europe. The ideas developed in the book took shape over several semesters spent both in Ann Arbor, Michigan, and in Tübingen, Federal Republic of Germany. I sincerely thank Professors H. Müller-Beck and U. Körber-Grohne as well as Drs. Helmut Schlichtherle, Ursula Maier, Stefanie Jacomet, and Manfred Rösch, all of whom generously provided opportunities to participate in field projects and to discuss problems of Mesolithic and Neolithic archaeology and paleoethnobotany—even though they may have disagreed with my theoretical approach.

On this side of the Atlantic, Drs. Karl Hutterer, Richard Ford, Robert Whallon, and V. Burton Barnes served on my dissertation committee at the University of Michigan. Thanks go to them for their encouragement to pursue the problem of forager-farmer interaction with a meagre data set. I would also like to express my appreciation to Dr. Michael A. Jochim, of the University of California at Santa Barbara, who introduced me to the problems of the Mesolithic/Neolithic transition in Southwest Germany. My thanks also go to Drs. Karl W. Butzer, H. Martin Wobst, Henry T. Wright, and William J. Parry, all of whom read and provided comments on earlier drafts of the manuscript; to Margaret Mahan, of the University of Chicago Press, who conducted the final editing of this volume; and last, but by no means least, to Carla M. Sinopoli, who read portions of several drafts and provided invaluable suggestions along the way.

Funding for the project came from the Rackham School of Graduate Studies at the University of Michigan, the National Science Foundation (Grant No. BNS 8506348), and the Wenner

Gren Foundation for Anthropological Research (Grant 4662). A scholarship initially to learn German was given by the German Academic Exchange Service (DAAD), and a Kontakt Stipendium provided by the DAAD supported my academic semesters at the Institut für Urgeschichte, Universität Tübingen. Final revisions of the manuscript were completed during my tenure as Visiting Scholar in the Center for Archaeological Investigations, at Southern Illinois University at Carbondale.

1

Introduction

Food production spread throughout the Old World over the course of several millennia, and farmers must have interacted with hunter-gatherers in different ways before cultivation and stockbreeding replaced hunting and gathering as the primary subsistence base. This book explores the possible nature and intensity of the relationship between hunter-gatherers and food producers during the fifth millennium b.c.[1] as agricultural communities were established in Central Europe.

During the fifth millennium b.c., a dichotomy could be drawn between coastal and interior hunter-gatherers in Central, Northern, and Western Europe. Coastal regions offered a rich and diverse set of resources, and in many areas the wealth of resources could be exploited by sedentary hunter-gatherers from a centrally located site. Resources were less abundant and more dispersed away from the coast: hunter-gatherers exploiting interior forests followed highly mobile strategies. This investigation explores the relationship between farmers and foragers (*sensu* Binford 1980) inhabiting the interior forests of Central Europe.

Research on the transition from foraging to farming in Central Europe has largely been founded on the assumption that the two subsistence strategies are fundamentally incompatible; consequently, foragers and farmers have been examined in isolation. The ethnographic literature, however, demonstrates that mixed strategies do exist. Foraging and farming can be combined in two basic ways. On the one hand, a single population may engage in hunting-fishing-gathering for part of the year and in cultivating-stockbreeding for the rest (Gross 1983). On the other,

---

[1]Unless otherwise specified, uncalibrated dates are used.

two populations may specialize in foraging and food producing respectively and then exchange goods and resources. Persistent, stable interactions of the latter type have occurred throughout the world in Africa (Bahuchet and Guillaume 1982; Blackburn 1982; Chang 1982; Harako 1976; Hart 1979; Hitchcock 1982; Tanno 1976), India (Fox 1969; Morris 1982; Sinha 1972; Smiley 1981), North America (Kelley 1955, 1986; Spielmann 1982, 1986; Wright 1967), and Southeast Asia (Gregg 1980; Hutterer 1976; Peterson 1978b; Pookajorn 1982). Cooperative forager-farmer interaction probably occurred at various times and places prehistorically, and I propose that it could have occurred in the fifth millennium b.c., when food producers expanded into Central Europe and encountered autochthonous foraging populations.

Investigating forager-farmer interaction calls for evaluating the problem from three points of view—anthropological, ecological, and archaeological. The first calls for determining when and under what circumstances two differently organized cultures can interact and yet maintain their own ethnic identity. The second requires identifying both the types of interaction that may have occurred and the conditions favoring the development of each type of relationship. The third entails determining whether interaction was feasible and did in fact occur.

In this book I develop a first-generation model examining all three aspects of forager-farmer interaction during the Mesolithic/Neolithic transition in Central Europe. My immediate goals are to (1) determine key social and economic variables shaping interaction between the two populations; (2) identify the types of relationships that may have existed; and (3) direct attention to the types of data that must be collected if prehistorians are to better understand the changes occurring as farming systems became established in Central Europe. Model building begins in Chapter 2 with a review of critical factors affecting the interaction of hunter-gatherer and cultivator-stockbreeder populations. An ecologically based model of population interaction is developed in Chapter 3. In Chapters 4 through 6 Early Neolithic subsistence strategies are examined, and ecological constraints that affected fifth millennium b.c. farmers are identified. In Chapter 7 the structure of wild resource exploitation by the foragers is considered in light of the ecological effects that cultivation and stockbreeding would have had on resource availability. Finally, Chapter 8 considers available archaeological data and discusses directions future research will have to take to better understand the relation between hunter-gatherers and cultivator-stockbreeders in prehistoric Europe.

Traditional Approaches to the Archaeological Problem

By the late sixth millennium b.c. farming villages were well established in the lower reaches of the Danube River. Between roughly 4600 and 4500 b.c., Neolithic farming villages appeared in the major river valleys throughout Germany, Austria, Czechoslovakia, Poland, and the Low Countries. The material culture, architectural style, and spatial organization of these Early Neolithic villages are virtually identical from Austria to the Low Countries. Because of the striking homogeneity of cultural remains, archaeologists attribute the appearance of these villages to a population movement (Childe 1929; Clark 1980; Dennell 1983; Quitta 1960; Starling 1985). This movement has been portrayed as a wave of farmers washing across Europe with a continually advancing frontier of interaction (Ammerman and Cavalli-Sforza 1973).

The wave model has shaped past investigations of the Mesolithic/Neolithic transition. In particular, it has caused forager-farmer interactions to be viewed as relatively short-term phenomena, persisting only as long as the initial colonizers entered into and passed through an area. Expulsion, acculturation, and avoidance are the types of interaction that have been commonly postulated. Each explanation is briefly considered below along with recent archaeological data substantiating or refuting the arguments.

Expulsion

Neolithic populations were thought to have grown so rapidly and to have had such a great need for fertile soil that forest clearance was believed to have destroyed the Mesolithic subsistence base (Clark 1980). As a result, Mesolithic populations were assumed to have been forced out of their original territories into "marginal areas," which were covered by soils too heavy to be worked with then extant Neolithic technology.

European archaeologists have recently begun to question assumptions critical to the argument. First, archaeological data now suggest the rate of population growth was low (Hammond 1981; Milisauskas 1977), and for a millennium following the initial explosive spread of settlements there was almost no expansion (Starling 1985:55). Second, botanical analyses (Willerding 1980, 1981, 1983) indicate soil fertility was not a problem. Moreover, with the exception of rare sites like Köln-Lindenthal (Buttler and Haberey 1936), Neolithic settlements tended to be small, dispersed villages with 4–10 houses spaced approximately 100 meters apart (Farrugia *et al.* 1973; Kuper *et al.* 1977; Lüning 1976). There is evidence that at least some villages were occupied continuously for up to 400 years, with a 25-year cycle of household construction, occupation, and abandonment. In some areas, a few isolated farmsteads seem to have been affiliated with every village (Kruk 1980). Thus, for at least some regions, small dispersed villages and isolated farmsteads appear to have formed the settlement pattern (Kruk 1980; Lüning 1982).

If the model of permanent villages with no expansion holds, the environmental impact of Neolithic farming needs to be reevaluated. Over long periods forests would have been affected within limited perimeters around the villages, but the effects of cultivation and stockbreeding would have been highly localized and of rather low magnitude. The traditional model, by contrast, has postulated large villages, rapid population growth with continuous settlement expansion, and village relocation every 20 to 30 years. A model of low-density dispersed villages with no settlement expansion allows for tracts of forest to have remained available for forager exploitation. The presence of villages and isolated farmsteads might have constrained forager mobility, but it would not necessarily have driven the hunter-gatherers into marginal areas.

Acculturation

Farming has been viewed as "more productive, more reliable, and providing richer sources of food" (Clark 1980:67) than foraging, and it was thought to have offered an inherently more attractive lifestyle. Archaeologists have generally assumed that Mesolithic foragers would therefore have adopted the technology, lifestyle, and social patterns of their farming neighbors (Menke 1978).

The validity of these assumptions can be challenged on two grounds. First, early Neolithic crops may not have been as reliable as they later became. Emmer, the most common wheat planted by early Neolithic farmers, is a grass native to arid lands. It is sensitive to soil types and excessive precipitation (Carleton 1901; Percival 1921; Peterson 1965). To some extent the effects of excessive precipitation can be offset by planting on well-drained soils, such as loess. But the precipitation levels of the Atlantic period are believed to have been very high (Fırbas 1949; Frenzel 1966; Iverson 1973; Lamb 1977). Even on well-drained soils, soil moisture may often have exceeded the toleration level of emmer wheat. Consequently, crop losses due to lodging and rusts can be expected to have occurred.

Second, whether an early Neolithic diet was "richer" than a Mesolithic one remains to be demonstrated. The early Neolithic subsistence economy is known to have included cereals, legumes, livestock, and wild foodstuffs, but there has not yet been a study of the degree to which these foods might have been available to the total population throughout the year. The predictability and harvest yield of Neolithic crops is beginning to be examined (Lüning and Meurers-Balke 1980), but paleopathological studies are lacking, and the effects of seasonal or periodic shortages on prehistoric populations have not yet been evaluated in detail. Moreover, the relative importance of plant and animal protein in the Neolithic diet has not yet been established, and the role that domestic livestock played in Neolithic economies is not fully understood. Some researchers argue that domestic animals were used primarily as a meat source (Bossneck *et al.* 1963; Higham 1966), while others (Bogucki 1982; Sherratt 1981) believe livestock was used for secondary products such as milk, blood, and raw materials like hair, wool, bone, and hide.

Until Neolithic subsistence strategies have been more fully examined, the argument that a Neolithic diet was richer (Clark 1980:67) can neither be substantiated nor refuted. Proposals about "acculturation" lack a compelling base.

Avoidance

Neolithic villages are restricted primarily to fertile, easily worked loess soils (Hammond 1981; Howell 1983; Kruk 1980; Sielmann 1971; Starling 1983), whereas foraging sites are located on both heavy glacial clays and light sandy soils (Gramsch 1971; Tringham 1968). The complementary distribution of sites associated with different subsistence strategies is usually viewed as reflecting two populations that avoid one another rather than interact (Cavalli-Sforza 1983; Clark 1980; Tringham 1968, 1971).

Recent research suggests that foragers and farmers may not have avoided one another. Neolithic ceramics have been found in "Mesolithic" contexts (Bagniewski 1981; Berlekamp 1977; Bicker 1933; Fansa 1985; Geupel 1981; Gramsch 1971; Taute 1966); Mesolithic stone and antler tools occur in Neolithic contexts (Gersbach 1956; Kind 1984; Uerpmann 1976); some open-air sites in southwest Germany (Hahn 1983; Taute 1974) show mixed lithic inventories, while Neolithic sites, such as Inzkofen near Munich (Födisch 1961), have lithic technologies that show clear affinities to the preceding Mesolithic and Paleolithic chipping techniques. Archaeologists have not yet developed satisfactory explanations to account for the mixed assemblages of both Mesolithic and Neolithic sites. One archaeologist suggests that "Neolithic" microliths at "Mesolithic" sites were probably "shot during a hunt and accidentally landed on a Mesolithic site"[2] (Taute 1974:77). Others tentatively suggest that Mesolithic populations selectively incorporated specific useful items, such as pottery, into their own lifestyle (Dennell 1983:175–176), but these explanations do not account for the presence of "Mesolithic" tools in "Neolithic" contexts.

Although the possibility of Neolithic farmers recycling Mesolithic tools has not been seriously evaluated, a few researchers are beginning to suggest a form of dynamic interaction. Most propose antagonistic relations, with the foragers raiding villages for livestock (Gramsch 1981; Bagniewski 1981). This argument is supported by well-fortified Neolithic villages (Fansa and Thieme 1985), but as early as 1953, Mázalek (1953) proposed a 2000 year co-existence between Mesolithic and Neolithic populations based on typological similarities in lithics. Some archaeologists (Taute 1974) discount Mázalek's arguments. Based on other data, Milisauskas (1978:91) has more recently proposed that hunter-gatherers may have traded game for grain, but he did not identify factors favoring such an exchange.

---

[2] All translations from German are by the author.

Interaction

Recent research undermines the assumptions on which expulsion, acculturation, and avoidance were predicated, while at least circumstantial evidence for interaction is increasing. As discussed above, the appearance of "Mesolithic" artifacts in "Neolithic" contexts and vice versa gives some indication of contact. Demonstrating long-term overlap of "Mesolithic" and "Neolithic" populations is difficult, if not impossible, with the currently available data. Palynologists have identified cereal pollen in very early contexts, and many support the possibility of an aceramic Neolithic (Kossack and Schmeidl 1974; Milojčić 1952; Müller 1947; Schütrumpf 1968).

The most serious problem is the difficulty in obtaining absolute dates from Mesolithic contexts. For one thing, organic remains have not survived at many Mesolithic sites. Such sites cannot be dated, and they could be contemporaneous with the Early Neolithic (Milisauskas 1978:80). For another, many of the known open air sites with mixed assemblages of Mesolithic and Neolithic lithics are surface scatters (see Hahn 1983) that cannot be dated absolutely. Other problems may arise from dates that do have good organic preservation. A review of dates reported in *Radiocarbon* suggests that archaeologists who, having submitted organic samples from good Mesolithic contexts and obtained dates that were "too young," readily dismiss the samples as being contaminated. They then rely on typological rather than chronometric techniques to date the sites. Certainly, contaminated samples should not be used; however, it is up to the archaeologists to identify such contamination before, not after, the dates are obtained. The practice of arbitrarily excluding dates that are too young and relying instead on cross-dating with typochronologies reifies the assumption that two dissimilar cultures cannot co-exist. Furthermore, the practice may be obscuring regional patterning of a chronometric overlap between the two strikingly different cultures.

The foragers and farmers during the mid-fifth millennium b.c. are viewed here as having different, but not necessarily mutually exclusive or incompatible, subsistence strategies and sociopolitical organizations. In fact, they may have developed equally viable alternatives for coping with the same environment. For this reason, the interaction must be considered within the temperate forest environment of mid-fifth millennium b.c. Central Europe.

Glaciers withdrew from Europe some five thousand years before the first farmers appeared, and, following the glacial retreat, the landscape slowly recovered from the effects of the ice age. First scrubby growth, then birch and pine forests, and finally deciduous forests spread throughout Central Europe in response to changing climatic and soil conditions. Assuming forest composition can be reconstructed from pollen and that the appearance of striking differences in pollen assemblages indicates a new forest composition, Firbas (1949) established a system of zones based on the relative frequencies of pollen from different tree species found in glacial and post-glacial strata. The Firbas zonation provides a standard when discussing environmental reconstructions in the Alpine Foreland, and most published pollen diagrams include a column showing the relationship between the new diagram and the standard Firbas zones. The pollen zones are significant archaeologically because they have provided and continue to provide a framework for dating occupation levels. Five post-glacial zones (Zones IV-VIII, Firbas 1949:49–51) are of particular interest to this study, for they present a sweeping overview of the slowly changing forest composition:

IV. Preboreal. Pine (*Pinus*) and birch (*Betula*) dominate, but hazel (*Corylus*), oak (*Quercus*), and elm (*Ulmus*) begin to appear.

V. Boreal. Hazel pollen increases sharply, then peaks and begins to decrease. Birch and pine are also relatively high. The Boreal can be further subdivided into an earlier period when pine was common and hazel was increasing, and a later period when both hazel and pine declined as pollen from the mixed oak forest increased. During this latter time, spruce (*Picea*) began to appear in the southeastern areas.

VI. Early Atlantic. The mixed oak forest (composed of oak, lime [*Tilia*, the European cousin of North American basswood], and elm) dominates, although hazel is still present. In the southeast spruce appears to be co-dominant with the mixed oak forest. Beech (*Fagus*), fir (*Abies*), and hornbeam (*Carpinus*) occur only sporadically, but a closed beech curve does appear in the southwest (Firbas 1949:231).

VII. Late Atlantic. The mixed oak forest continues to dominate, but elm and lime diminish while ash (*Fraxinus*) increases. Hazel expands towards the end of the period. Beech and fir occur sporadically; although they are frequent in the southwest.

VIII. Subboreal. Fir and, particularly, beech expand dramatically while the mixed oak forest retreats. Hazel forms a secondary peak during the early phase of the Subboreal. Cereal pollen is common.

Foragers inhabited Central Europe during the Preboreal, the Boreal, and the Atlantic. The first farming villages were established in the mid-fifth millennium b.c., when the Atlantic period was in full swing. The climate was generally warm with high levels of precipitation, hazel had retreated, and mixed oak forests had become established throughout Central Europe. But forest composition continued to change. Throughout the Atlantic period (Firbas Zones VI and VII) spruce spread slowly north-northwest from Austria, while fir spread north-northeast from Switzerland. Furthermore, beech forests obtained a toehold north of the Alps during the early Atlantic, and by the end they had spread northwards throughout Central Europe. Indeed, the sharp increase of beech in the pollen profiles marks the end of the Atlantic and the beginning of the Subboreal.

The establishment of beech forests must have had serious consequences for local human, plant, and animal populations. The canopy of an oak forest is relatively open and allows large amounts of sunlight to reach the forest floor. An exuberant undergrowth of mixed shrubs, forbs, and grasses develops, and the diversity of plants supports a variety of wildlife. In contrast, the canopy of a beech forest is closed and the forest floor is heavily shaded. Other than a flush of spring annuals prior to the emergence of the leaves, only shade-tolerant sedges, ferns, and a few grasses are found.

Early Neolithic cultivator-stockbreeders arrived in some parts of southern Central Europe during the transition between the early and late Atlantic (Firbas Zones VI and VII), just as beech was beginning to spread but before it dominated the forests. Coming as they did in the earliest stages of a major change in forest composition, farmers must have compounded the magnitude of change. The farmers brought with them a suite of domestic crops and livestock that presumably fulfilled most of the basic dietary requirements. However, the transfer of crops and cultivation techniques into a new environment initially may have led to a lowering of reliability, and it may have been necessary to supplement cultivation and stockbreeding with wild resources. At the same time, Neolithic villages would have offered a concentration of resources not only in the form of domestic crops and livestock, but also in the form of wild plants and animals, whose proliferation would have been encouraged by the creation of forest clearings near a village. Instead of viewing the activities of the food producers as degrading the climax forest, it may be more appropriate to view their activities as adding an element of diversity to it.

Both indigenous and immigrant populations must have adapted their subsistence strategies to resources found in the changing composition of the deciduous forests. The relationship between hunter-gatherers and cultivator-stockbreeders during the Atlantic period, therefore, might profitably be examined not as the replacement of one population by another, but as the mutual adaptation of both to changing resource configurations. Thus a fourth possibility, that of mutual cooperation and co-existence, must be added to the list of acculturation, expulsion, and avoidance. The model developed here is intended to examine under what conditions cooperative co-existence would arise and to identify the types of data needed to determine whether or not such an interaction did in fact develop.

The Alpine Foreland in the Sixth Millennium b.c.: the state of current evidence

The archaeological data used in developing the model come from the Alpine Foreland of southern Germany. The Alpine Foreland curves from southwest to northeast between the Alps and Jura. It runs across central Switzerland, through southern Germany, and into Upper Austria. Pleistocene glaciers moved over the region repeatedly; consequently the Alpine Foreland is covered by sheets of boulder clay, outwash plains, and pockets of loess; it is dotted with glacial lakes, drumlins, and end moraines. Three major European rivers, the Rhône, the Rhine, and the Danube, drain the area. The primary focus of this study (Fig. 1) is southwest Germany, where the Alpine Foreland is delimited to the south by Lake Constance (Bodensee), and to the north by both the Swabian Alb (the northern extension of the Jura) and the Danube River.

Archaeologically, the Alpine Foreland is well known for waterlogged Middle Neolithic and Bronze Age villages, which have been investigated sporadically over the past century (Reinerth 1936;

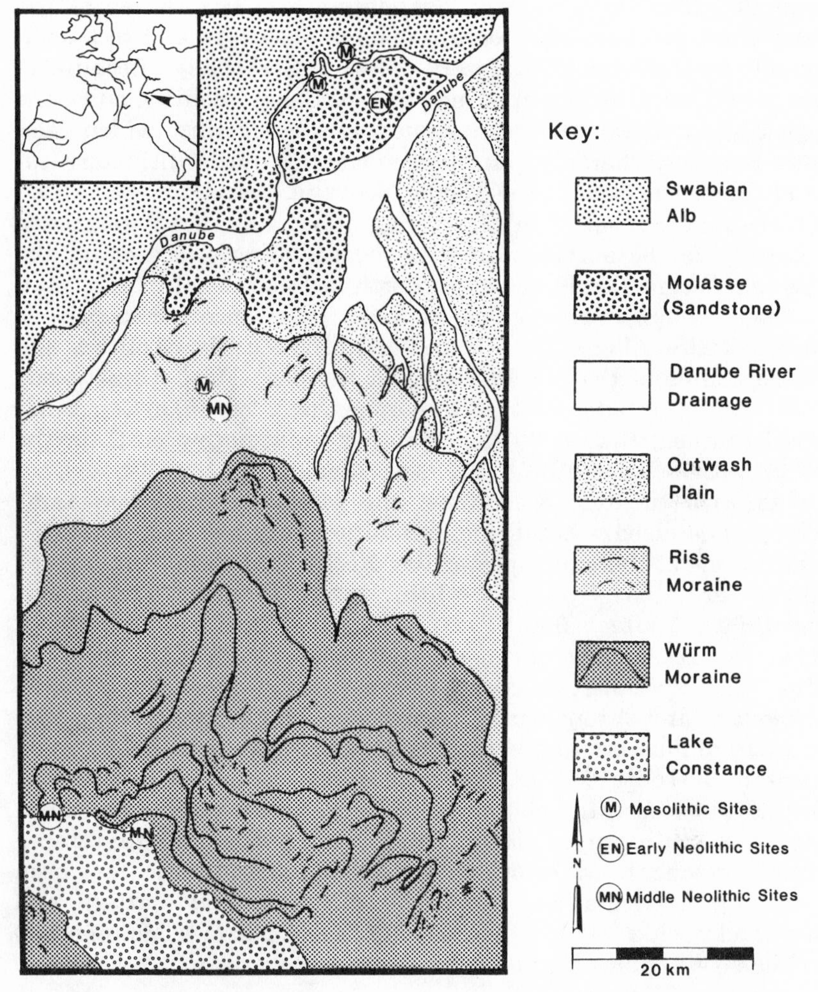

Figure 1. Study Area

Schmidt 1984; Schmidt 1936; Schlichtherle 1984), as well as the Paleolithic and Mesolithic occupations of the caves and rock shelters of the Swabian Alb (Peters 1934; Taute 1980). Because of the heavy soils and the apparent lack of Early Neolithic sites, the region has traditionally been viewed as a refuge into which remnant hunter-gatherer populations retreated until cultivators expanded into the area in the Middle Neolithic period.

Examination of Neolithic occupation has been restricted primarily to the spectacular waterlogged villages located on the Riss and Würm glacial moraines north of Lake Constance, as well as along the northern shore of Lake Constance (Schlichtherle 1984). These villages date to the end of the Early Neolithic and the beginning of the Middle Neolithic periods. Most recently, one Early Neolithic site west of the upper arm of Lake Constance (Aufdermauer *et al.* 1986) and another immediately north of the Danube River (Kind 1986) were excavated.

As with the rest of Europe, the Mesolithic period remains poorly understood. Mesolithic excavations are not as frequent as excavations at the more spectacular Neolithic, Bronze, or Iron Age villages or in the Upper Paleolithic levels of caves. Throughout the 1960's Taute (1966, 1967, 1980) conducted a program of Mesolithic research in the caves and rock shelters of the Swabian Alb. The specialist analyses have been published, but the typological and chronological synthesis of Taute's excavations and analyses of museum and private collections have not. Research on open-air Mesolithic sites has been limited, but investigations around the Mesolithic shoreline of the Federsee are increasing in pace. Jochim (in press; Jochim and Gregg 1984) has excavated a multi-component waterlogged site, Schmidt (1984) has been analyzing materials from known Mesolithic surface scatters, and the Landesdenkmalamt (the State Office of Historic Monuments) in Baden-Württemberg has initiated a new program of systematic excavations (Wagner 1986).

Palynologists were the first to suggest that cultivators moved into the Alpine Foreland of southwest Germany in the Early Neolithic. In 1947 Müller analyzed pollen profiles from the Federsee (the largest lake between Lake Constance and the Danube River) and from the western end of Lake Constance. In both profiles she identified the presence of cereal pollen from strata that Firbas later assigned to the interface of Zones VI and VII (Müller 1947:76, 80; Firbas 1949:368, 370). Schütrumpf (1968) reevaluated the chronological placement of Müller's samples and argued for a later dating. Schütrumpf's arguments are based on using percentages of beech in the pollen profiles as a basis for relative dating. This method of obtaining relative dates may be suitable for sites with the same environmental and geomorphological conditions; however, the sites Schütrumpf compares lie 40 km apart on different soils and in different microclimates.

The environmental differences alone could account for differing forest compositions, which could lead to a spurious chronological alignment. Furthermore, Schütrumpf does not satisfactorily explain the appearance of cereal pollen in strata well below the Middle Neolithic villages he examined. Müller's conclusions, although still open to considerable debate, have not been successfully invalidated. Moreover, cereal-type pollen has also been found in early Atlantic (Firbas Zone VI) levels in profiles from two locations in the Bavarian region of the Alpine Foreland (Kossack and Schmeidl 1974). The Bavarian findings support Müller's argument that cereal cultivation was introduced to the Alpine Foreland in the early Atlantic.

Archaeological data provide more concrete evidence of interaction. Three sites outside of the study area are of particular importance. Inzkofen, near Freising in Bavaria (Födisch 1961) is an open-air campsite on the northern border of the Alpine Foreland. The site has Epipaleolithic stone tools as well as Neolithic ceramics and polished stone tools. Sarching, a multi-component open-air Mesolithic site in Bavaria (Schönweiss and Werner 1974), was inhabited at the same time that a neighboring Early Neolithic village was occupied (Menke 1978:37). Similarily, Birsmatten-Basigrotte, a multi-component Mesolithic cave site in Switzerland, was occupied throughout the Atlantic period—even when Early Neolithic villages were established in the neighboring valley (Menke 1978:47), and a Middle Neolithic ceramic sherd was found in the uppermost Mesolithic occupation levels (Bandi 1963:249).

Within the study area, evidence for interaction comes from rock shelters and caves in the Swabian Alb (Hahn *et al.* 1973). Taute (1980: Table 1) indicates that Falkenstein, Bettelküche, Burghöhle Dietfurt, Zigeunerfels, Lautereck, and Schrägewand have mixed Mesolithic and Neolithic deposits. Of these sites, only Lautereck has been published (Taute 1966). It provides evidence of interaction between Mesolithic and Neolithic populations.

Lautereck, a rock shelter on a tributary of the Danube, has five levels. The second through fourth levels (Cultural Layers B–D) contain Neolithic ceramics along with lithic materials, while the Terminal Mesolithic level (Cultural Layer E) has a tool industry comparable to industries found at Early Neolithic sites. Neither domestic cereals nor faunal materials were recovered; however, this may be a function of excavation techniques, for soil was neither sieved nor floated. Despite the lack of domesticated remains, Taute argued that the lowest component contained a local Terminal Mesolithic assemblage showing a definite relationship with the fully Neolithic groups (1966:502). Fish remains occurred frequently in the Neolithic as well as the Mesolithic components, and Taute considers the rock shelter to have been repeatedly used as a fishing camp. Burghöhle is another cave with well defined Mesolithic and Neolithic occupations, and with quantities of fish remains (Brunnacker *et al.* 1979).

Finally, a number of Alb caves well north of the Danube and outside of the study area have Mesolithic and Neolithic components. Many of these later sites were excavated during the late 1890's or early 1900's using excavation techniques that were not as precise as those employed today. The descriptive accounts of these excavations emphasize the Upper Paleolithic levels, and Neolithic materials are discussed only in passing (Riek 1934). Nonetheless, Taute (Hahn *et al.* 1973:149–154) interprets the repeated occurrence of these materials as evidence of remnant populations of foragers who continued to exist throughout the Neolithic period.

Early Neolithic materials other than those found in caves and rock shelters were limited to isolated finds, but by the late 1970's some enigmatic surface collections containing mixtures of Mesolithic and Neolithic materials began to be classified as belonging to the transitional phase. In the late 1970's a cluster of Early Neolithic settlements was found (Heiligmann 1983) within a few kilometers of a concentration of Mesolithic sites (including Lautereck). One of the Early Neolithic villages, Ulm-Eggingen has been excavated, and Mesolithic microliths have been recovered at the site (Kind 1986). The discovery of the Early Neolithic sites suggests Mesolithic and Neolithic populations may have overlapped. Mesolithic materials at the Early Neolithic village and Early Neolithic ceramics in Mesolithic contexts suggest both contact and interaction between the two populations may have occurred.

Examining the relationship between indigenous hunter-gatherers and immigrant farmers is hampered by several problems. First, the archaeological data base is fragmentary and uneven. Although Mesolithic and Neolithic research has been continuing for several decades, early researchers have not published their findings. What information is available exists primarily as brief preliminary reports from the early 1900's (Reinerth 1928, 1936); they contain insufficient detail for a study of population interaction. Moreover, traditional research has been directed toward establishing typologies and resolving chronological problems. These studies establish temporal frameworks for the Mesolithic and Neolithic periods but their underlying assumption is that typological differences are synonymous with cultural differences, and that the two cultures *ipso facto* could not coexist. Thus, traditional research has been oriented toward examining the two populations as separate, independent, and chronologically discrete. With such a philosophical foundation, research has not been directed toward obtaining data needed to examine the possibility of interaction between two differently-organized populations pursuing different subsistence strategies within a single homogeneous region.

The problem is common in archaeological research, and it is unique neither to European archaeology nor to the Mesolithic/Neolithic transition. Archaeology lacks a theoretical foundation for examining long-term interaction between two differently organized populations. Satisfactory data collection is unlikely to occur before theoretical models of population interaction are developed. Once such models exist, the extant data base can be evaluated and future research more specifically directed toward obtaining the necessary information.

Methods

Interaction between two differently-organized social groups does not inevitably lead to the acculturation of one, and cultural differences can persist in spite of group contact and interdependence. Barth (1956; 1969:18) argues that stable, persistent, and vitally important social relations can be maintained between ethnic groups. He suggests three types of interaction (1969:19). First, each population may occupy different niches and have minimal competition for resources. In such a situation, articulation is limited primarily to trade or ritual interaction. Second, each may inhabit a separate territory and exploit resources within their respective territories. Interaction between such groups would be limited to border maintenance. Third, each may exploit basically different sets of resources within the same region and provide one another with important goods and services. This book develops a model exploring the last type of interaction.

Ecological models, closely related to those of classical microeconomics have proved useful in examining decision making and resource exploitation. These models have been successfully applied to investigations of hunter-gatherers (Keene 1981; Winterhalder and Smith 1981), tropical horticulturalists (Keegan 1986), and forager-farmer interactions in North America (Spielmann 1982, 1986). They will be used here in examining whether or not a cooperative interaction could have developed between Mesolithic foragers and Neolithic farmers. The evaluation will be based on the costs of filling the caloric and protein requirements of each population. Recommended daily intakes of energy as suggested by the World Health Organization (WHO 1974: Table 1) are presented in Table 1. The populations considered will be a reference band of hunter-gatherers and a reference village of cultivator-stockbreeders. Both will be arbitrarily generated in Chapter 6 using a stochastic procedure in conjunction with demographic tables published by Weiss (1973).

The regularity and predictability of subsistence resources emerge as the key variables in long-term interactions (see Spielmann 1982, 1986 for a complete discussion). As a first step, detailed models must be developed of both forager and farmer subsistence strategies to account for variability in the supply and demand of critical resources. Such models are achieved in part by adapting existing optimal foraging models (Jochim 1976; Keene 1982) to the Terminal Mesolithic period; however, the Neolithic farming economy is poorly understood. The relative importance of domestic and wild foods in the Neolithic diet has not been established, and constraints affecting harvest yields have not been clearly delimited. It has therefore been necessary to develop an optimal model of temperate forest wheat farming in which the likely variability in harvest yields, planting strategies, and scheduling conflicts are identified. Based on this optimal farming model, the amount of domestic and wild resources in the diet; the extent of land needed to support a Neolithic village; and the

Table 1

Recommended Protein and Energy Intakes

Age	Energy		Protein (grams)
	kilo-calories	mega-joules	
Children			
<1	820	3.4	14
1–3	1,360	5.7	16
4–6	1,830	7.6	20
7–9	2,190	9.2	25
Male adolescents			
10–12	2,600	10.9	30
13–15	2,900	12.1	37
16–19	3,070	12.8	38
Female adolescents			
10–12	2,350	9.8	29
13–15	2,490	10.4	31
16–19	2,310	9.7	30
Adult males			
(moderately active)	3,000	12.6	37
(very active)	3,500		
Adult females			
(moderately active)	2,200	9.2	29
(very active)	2,600		
Pregnant females			
(second half)	+350	+1.5	38
Lactating Females			
(first 6 mos.)	+550	+2.3	46

Source: WHO 1974: Table 1

regularity with which the farmers might have provided the foragers with domestic foods can be calculated.

Using models to illuminate aspects of prehistoric societies is not new, and European archaeologists have devoted much time and energy to developing and examining physical models under the rubric of experimental archaeology. In particular, working prehistoric farmsteads (Reynolds 1984), house and village reconstructions (Mausch and Ziessow 1985; Meyer-Christian 1976; Startin 1978), actual living situations, and the experimental cultivation of emmer and einkorn wheat (Lüning and Meurers-Balke 1980; Steensburg 1979) have provided insights into prehistoric subsistence systems. The purpose of these experiments has not been to imitate or perfectly replicate prehistoric systems. Rather "there exists throughout the hope that out of practical experience, new hypotheses can be developed" (Lüning and Meurers-Balke 1980:309).

My goal is the same, but the simulation of the Neolithic subsistence economy differs somewhat from the actualized models preferred by European archaeologists. First, the simulation is abstract. Second, ecological theory and fundamental concepts of population biology along with historical crop records from the region provide parameters for establishing the reliability and predictability of Early Neolithic food production systems. Third, I used a computer to calculate, track, and record the effects of the frequency and severity of harvest fluctuations. The value in developing a computer simulation is that a machine performs the necessary calculations in a fraction of the time that would be required were the model to be executed by hand. Consequently, the model can consider several alternatives to a complex set of interrelated variables. Aside from mechanical aspects concerning the number and complexity of variables considered, computer simulations differ little from the construction of any model. Identifying the critical parameters and specifying their relationships are the most important aspect of the exercise. Like physical models normally favored by European archaeologists, simulations are not intended to provide a precise reconstruction: rather, their purpose is to generate testable propositions.

Scope of Analysis

The simulation of Neolithic subsistence strategies is a dynamic model, but I have developed it in order to predict an optimal farming strategy. I then use this strategy to develop a *static* model of forager-farmer interaction. There are several drawbacks in developing a static, rather than dynamic, model of forager-farmer interactions. A static model assumes an optimal or average interaction, which in reality was probably an uncommon occurrence.

The primary benefit in developing a static model is that it allows the critical variables and the relationships among them to be identified, and it generates testable propositions through which the model can be evaluated and refined.

Furthermore, a dynamic model requires relatively detailed information on the annual fluctuations of resource availability. Because such data are not currently available for Neolithic Europe, the development of a dynamic model at this point would entail making a series of *ad hoc* assumptions. At the level of first-generation modeling, it is sufficient to develop a simple model that subsequently can be refined.

My goal is to develop a model examining the constraints that could have affected the nature and intensity of the interaction between indigenous Mesolithic foraging and immigrant Neolithic farming populations during the fifth millennium b.c. in Southwest Germany. Whether such a relationship actually obtained must be determined from the archaeological record. The available data base is scanty, and much remains unpublished. Thus, hypotheses derived from the model can be evaluated in only a very general sense; moreover, all the available data suggest that overlap occurred between the Mesolithic and Neolithic populations and that cooperative interaction may have occurred. It would be inappropriate to turn around and test the model with the very data used to create it. Independent archaeological tests of the model developed here will have to be completed in the future. The logic underlying the model and the selection of parameters are, however, subject to evaluation at all stages. The strength of the model must for the present be based on the logic of its construction.

2

Mobility, Subsistence, and Social Organization

One assumption underlying this inquiry is that a congruence exists between a group's resource base, extractive technology, mobility strategies, and socio-political organization. Each aspect places restrictions on the range within which the others operate, but no single aspect determines any other; instead, a range of variability is to be expected. Therefore, before considering the interaction between Mesolithic and Neolithic populations it is necessary to identify factors that may have affected the relationship. This chapter examines ways in which mobility strategies and resource procurement affect the organization of small-scale egalitarian societies, and it identifies the range of variability that might be expected with both foragers and farmers. First, similarities and differences in the social organization of hunter-gatherers and horticulturalists are evaluated; then a construct of cultivation systems is developed. Finally, cultivation strategies are examined with particular reference to their effect on hunting-gathering populations.

Settlement Permanence and Social Organization

The degree of settlement permanence critically influences social organization. The longer populations live together, the more important social mechanisms become in moderating social interactions. Binford (1980) began to develop an overarching model of mobility strategies for hunter-gatherers by distinguishing between *residential* and *logistical* mobility. The former concerns the movement of an entire residential group, while the latter notes the movement of individuals or task-oriented groups out from and back to a residential location. According to Binford's construct, residentially organized hunter-gatherers move as a unit to the available resources; whereas, logistically organized hunter-gatherers send special task groups out to collect the resources and bring them back to the group. Binford argues that *residentially* and *logistically* organized hunter-gatherers, which he respectively identifies as foragers and collectors, are polar types on a continuum of mobility strategies.

Eder (1984:848), on the other hand, suggests mobility and settlement permanence exist at two different levels of analysis. While agreeing that settlement permanence is determined by the frequency with which an entire group relocates, Eder argues that settlement mobility is a group phenomenon and should be investigated at the group level. Eder also concurs that subsets of the local group (such as task groups, single households, or groups of households within the local group) move about to exploit resources; but he suggests this type of mobility is best examined at a subgroup level. According to Eder's construct, settlement permanence and group mobility are separate and not necessarily related phenomena; any settlement or camp, regardless of its degree of permanence, can act as a base from which smaller groups may operate. Thus, residential and logistical mobility are not continua of the same phenomenon. Instead, they are separate dimensions that cross-cut one another.

Following Eder's arguments, a general construct of population mobility must encompass groups with (1) low residential mobility and high logistical mobility; (2) low residential mobility and low logistical mobility; (3) high residential mobility and low logistical mobility; and (4) high residential mobility and high logistical mobility. With this general construct of population mobility in mind, similarities and differences in the organizational structure of small-scale, egalitarian societies may be attributed to an interaction of food extraction technologies and two-dimensional strategies of population mobility.

In the following discussion, one model of socio-political organization is constructed for small-scale, egalitarian societies that are residentially mobile populations, and another is developed for residentially sedentary populations. Rather than coining new phrases, these populations be referred to respectively as "bands" and "tribes." The two models subsequently provide a foundation for examining aspects of the interaction between hunting-gathering and food-producing populations during the Mesolithic/Neolithic transition.

Band Societies

The "Man the Hunter" symposium in 1966 marked a juncture in the study of hunter-gatherer societies. Until the symposium, a "pristine" band had been thought to be a patrilineal, exogamous group with virilocal residence subsisting primarily on hunted game (Service 1962). It provided a tidy model. Males were believed to have resided in the same area throughout their lives, and it was assumed that as children they learned how to exploit their territory efficiently. Women were thought to have been exchanged, thereby providing affinal ties with other bands (Owen 1965).

Symposium participants reformulated the model by making two assumptions: "(1) that hunter-gatherers live in small groups and (2) that they move around a lot" (Lee and DeVore 1968:11). From these assumptions five characteristics of hunter-gatherer societies were derived: (a) property is not accumulated, because hunter-gatherers move frequently; (b) low population densities are maintained by small groups occupying relatively large territories; (c) local groups do not have exclusive rights to resources; (d) food surpluses are neither collected nor maintained; (e) mobility prevents groups from becoming strongly attached to any single area (Lee and DeVore 1968:12). The reformulated model marked a turning point in the examination and explanation of hunting-gathering societies, but it is applicable only in the case of hunter-gatherers living in small groups with high residential mobility. Such hunter-gatherers move through the landscape in ephemeral bands usually consisting of one or more extended family groups. Bands tend to be organized along consanguineal and affinal kinship lines, with specific membership fluctuating constantly. The groups are neither organized nor recognized as corporate social entities within a larger social unit, although bands tend to recognize one another according to the geographical territory they habitually exploit (Needham 1962). Ownership of and access to resources is held in common by the group (Leacock and Lee 1982:4).

Residentially mobile populations generally do not store foodstuffs against seasonal or unexpected lows in resource availability. Instead, mobility itself provides the key adaptive mechanism in buffering against periodic resource stress. When lows occur, groups disband and members disperse along affinal or consanguineal kin lines to other regions. Flexible, ego-centered kin networks thus ensure complementary access to a variety of resource territories that may lie within different environmental regimes. Mobility also acts as a mechanism for resolving conflicts. When personal conflicts arise, one party simply leaves to join another group.

A high level of residential mobility is the most significant aspect of band societies. Economically, it buffers against periodic lows in resource availability and obviates the need for food storage. Socially, it furnishes a mechanism for conflict resolution and offers a means of maintaining social ties within a diffuse population. Residential mobility becomes a less viable option for obtaining access to resources, for resolving personal conflicts, and for maintaining social networks when the freedom of movement is circumscribed for whatever reason. As group mobility decreases, the "loose non-corporate nature of the small-scale society cannot be maintained" (Lee and DeVore 1968:12); and social mechanisms must be employed to resolve personal disagreements and to offset resource fluctuations.

The band model that emerged from the "Man the Hunter" symposium concerns the social organization of hunter-gatherers with high residential mobility. It is not applicable to hunter-gatherers who have permanent or semi-permanent settlements, accumulate property, and have relatively high population densities, and who may also have exclusive rights to some resources and may both collect and maintain food surpluses (see Koyama and Thomas 1979). They do not meet the criteria allowing them to be classed as band-level societies. It is therefore necessary to consider how sedentary hunter-gatherers might be organized and the ways in which their organization differs from the band-level organization of non-sedentary hunter-gatherers.

Generic Tribes

Sedentary, non-hierarchical small-scale societies currently pose a problem to anthropologists. The term "tribe" is used generically to refer to corporate, non-stratified residential groups composed of several nuclear families living together more-or-less permanently and exploiting the local environment. Such residential groups are organized on the basis of primary kinship ties and usually lack institutionalized political offices, full-time craft specialization, and inherited wealth or prestige statuses (Braun

1977:80–81). Regionally, social interaction is maintained via secondary kinship ties as well as through a web of personal ties, trade relations, marriage alliances, and institutionalized secret societies, associations, and sodalities (see Service 1962:102). Despite the range of variation in tribal societies, a suite of features recur (Fried 1975; Sahlins 1968, 1972; Service 1962) including increased residential permanence resulting in greater face-to-face contact, unilineal kin reckoning, defined group boundaries but fluid membership, and production determined by the nuclear family. Tribal societies thus appear to operate under a limited set of general rules that are adaptable to a variety of situations and have resulted in an astonishing array of seemingly incomparable social forms.

Sedentary populations encounter a number of problems that non-sedentary populations resolve simply by moving. The most serious are resource exploitation, allocation, and consumption. To avoid resource depletion, access to resources must be limited. One concept to be expected in sedentary—but not mobile—egalitarian, small-scale societies is exclusionary use-rights to resources. Exclusionary use occurs when only a limited set of individuals may exploit the available resources. It occurs at both regional and local levels. Regionally, local groups hold vested rights in specific territories, and, within each local group, individuals (or households acting as economic units) hold legally sanctioned rights to exploit specific resources. Exclusionary use-rights should not be equated with the exclusive ownership of or right-of-way through specific properties or territories. Moreover, use-rights need not be physically defended. Boundaries, such as geographical landmarks, topographical features or symbolic markers in the form of economically valuable trees, fences, or borders of stones, may suffice to proscribe use. Exclusionary use-rights simultaneously protect resources from being over-exploited or exhausted and limit the amount of conflict both within and between social groups.

Exclusionary use-rights are absent in bands (see Kelly 1985; Leacock and Lee 1982) but they appear to be common in tribal societies. Sahlins suggests (1972:41) that within tribal societies each family has usufructory rights to productive resources, even though actual ownership of the resource may lie with the lineage or village. In unilineal systems, individuals retain legal access to the cross lineage's resources (Brookfield and Brown 1963), and during periods of extreme duress the cross lineage can be looked to for help. Thus, in tribal societies kinship becomes a vehicle for excluding whole segments of the population from exploiting particular resources. Conversely, kinship defines and ensures access to specific productive resources both in the residential village and in neighboring villages.

Just as high residential mobility is a significant characteristic of band societies, sedentariness is a critical aspect of tribal societies. Extremely low levels of residential mobility result in the following key characteristics of tribal societies: (a) moves are infrequent, and property can be accumulated; (b) face-to-face contacts occur for extended periods, and social mechanisms are needed to resolve conflicts; (c) unless somehow conserved, resources will be over-exploited; therefore, the exploitation of specific resources may be vested in individual households; (d) food may be stored during one part of the year for later allocation and consumption.

Horticulture and Sedentism

Horticulture is often seen as a prerequisite of tribal societies; however, it is the low residential mobility rather than cultivation *per se* that is responsible for organizational characteristics of tribal societies. The organizational structure of horticulturalists derives from two pragmatic aspects of crop cultivation. First, crops are a stationary resource. Since certain activities can be undertaken only at specific times, crops tether populations to a specific location: populations must return minimally at the planting and harvest seasons, as well as sporadically to recover stored produce. The life cycle of the plant in conjunction with the seasonal climatic regime constrains group residential mobility. Second, the production, storage, and consumption of crops entails a long-term commitment to one particular form of subsistence activity. Planning must be undertaken to determine the amount of foodstuffs desired at the end of the production year, labor is invested throughout the production cycle, and the allocation of stored resources is controlled during the consumption year.

In the resulting subsistence system the consumption phase of one cycle, the production phase of a second, and the planning of a third may overlap. Adopting cultivation as a food procurement strategy has several implications for the social organization of the population. Mechanisms must previously exist or be developed to

(1) ensure the production, storage, and allocation of the crops as well as to conserve seed for the next season's crop; (2) mediate internal and external conflicts and settle disputes; and (3) foster and sustain group interaction.

Like horticulturalists, sedentary hunter-gatherers have relatively dense permanent residential groups with at least semipermanent population aggregation. Aggregation is made possible either by predictable resources available throughout the year or by seasonal surpluses managed in conjunction with a technology that allows resources to be stored for future use. Subsistence activities are usually performed by nuclear families, but cooperative activities integrate the group. To compensate for quantitative and qualitative imbalances of local resources, networks based on an individual's affinal, collateral, and fictive kin ties are commonly employed to obtain food and non-food resources from other regions in exchange for local goods. In contrast, hunter-gatherers with high residential mobility are loosely organized around egocentered kin lines. They follow a pattern of aggregation and dispersion in response to the seasonal availability of resources. They might not have a predictable, year-round, highly productive set of resources; they might lack either the resources or the technology that would allow them to store foodstuffs, or they might have both but choose not to use them.

Hunter-Gatherer Responses to Horticultural Systems

Three critical factors affect the spread of food production: the social organization of the indigenous hunter-gatherer populations; the mobility strategy of these populations; and whether food production was introduced by a population of cultivators moving into an area or by the transmission of cultivation without an attendant population movement. These factors combine to form four different situations: (1) residentially mobile, band-level hunter-gatherers incorporating cultivation into their annual round; (2) tribal-level, sedentary hunter-gatherers adopting cultivation; (3) cultivators moving into an area occupied by residentially mobile hunter-gatherers; and (4) cultivators moving into an area occupied by sedentary hunter-gatherers.

In much of Europe and most of North America the use of cultigens spread without an attendant population movement. Recent research suggests the adoption of food production by recipient hunter-gatherer systems occurs when, but not before, the hunter-gatherers have developed a storage technology and social mechanisms for both controlling the allocation of stored resources and ensuring the conservation of seed for subsequent crops (Wills 1985). The spread of food production systems into Central Europe, in contrast, entailed the movement of cultivators into an area inhabited by hunter-gatherers. Although some indigenous hunter-gatherers may have turned to cultivation, they could have incorporated the use of cultigens in their subsistence strategies without actually becoming farmers themselves.

The initial acceptance of immigrant cultivators would vary with the social organization and residential mobility of the indigenous hunter-gatherers. If the hunter-gatherer population is sedentary, it could be expected to have a tribal-level social organization. This has two ramifications. First, cultivators moving into an area would infringe upon previously allocated territories, and direct competition could develop between the tribally organized, sedentary hunter-gatherers and the immigrant cultivators. Second, the arrival of immigrant cultivators might set a chain of events in motion. Down-the-line bumping might occur, with one local group nudging their neighbors. The net effect would be a regional realignment of territories. One consequence of such a realignment might be smaller per-capita territories, and this in turn could increase the need to enhance local productivity. Adopting cultivation would be one solution.

If concepts of exclusionary use-rights, resource enhancement, and food storage were already present in the sedentary hunting-gathering population, adopting cereal cultivation would require only a shift of emphasis in the degree of resource enhancement rather than a complete change of subsistence strategies. But it would still require learning a complex, new technology.

Other results might be expected in an encounter between cultivators and hunter-gatherers with high residential mobility. The cultivators might effectively prevent the hunter-gatherers from exploiting specific resources as they had previously used them, but simultaneously the concentration of domestic resources might itself represent new opportunities for the hunter-gatherers. One response might be an increase in hunter-gatherer mobility, with a realignment of the annual subsistence round. The restructuring of subsistence activities could entail incorporating the cultivators as a resource in the subsistence strategy of the foragers. If the foragers did incorporate aspects of domestic foodstuffs into their annual cycle, then some sort of cooperative or competitive interaction with the farmers must be expected. The nature and intensity of the interaction would have been affected by the degree to which the Neolithic populations influenced the resource availability within the deciduous forest.

Adopting farming itself or incorporating the cultivators as a potential resource in a revamped subsistence strategy could have occurred in different areas of Central Europe. During the fifth millennium b.c. sedentary hunter-gatherers in northern Europe subsisted on the abundant resources found in coastal areas. The spread of Early Neolithic farming communities stopped just south of this area, and for more than a millennium the sedentary hunter-gatherers did not adopt cultivation themselves, while simultaneously preventing cultivators from moving into the area (Fansa 1985). Only when critical marine resources were no longer available, did the indigenous sedentary hunter-gatherers turn to cultivation and stockbreeding (Rowley-Conway 1984).

In contrast, Central Europe during the fifth millennium b.c. appears to have been inhabited primarily by residentially mobile hunter-gatherers (see Jochim 1976; Müller-Beck 1983). Population densities were low, and the immigrant cultivators established villages on terraces along the major rivers. The degree to which the cultivators affected indigenous hunter-gatherers would have depended on the land use patterns of the cultivators.

Land Use and Residential Mobility Among Shifting Cultivators

It is tempting to draw parallels between the rapid spread of wheat farming in Neolithic Europe and in historic North America. There are, however, critical differences. The introduction of wheat farming into and subsequent spread throughout the New World was stimulated in part by a growing global economy. Colonies were initially founded and financially supported by European companies with the express intent of providing Europe with much needed furs and raw materials for their manufacturing centers and an outlet for their finished goods. The spread of European populations throughout North America was subsequently stimulated, at least partially, by the need to provide eastern industrial centers with food and raw materials. The destruction of the environment and depletion of the wild game throughout North

America was primarily a function of the scale of the population movement, the insatiable demands of eastern industrial cities, and trade relations with Europe.

The fifth millennium b.c. spread of farming populations into Central Europe was not on the same scale as the migration of Europeans into and across North America during the seventeenth and eighteenth centuries A.D. First, Neolithic domestic crops were limited to wheat, peas, lentils, flax, and occasionally poppy. Barley rarely appears at Early Neolithic sites, and it is not thought to have been an important crop. In addition to dogs, domestic livestock was limited to cattle, goats, pigs, and sheep. Horses had not yet been domesticated; it would be another millennium before either ducks or cats were domesticated and several millennia before chickens were introduced to Europe. There is no evidence for either the wheel or the plow, although both probably existed from the Middle Neolithic on. Nor is there evidence to suggest that Early Neolithic populations were connected to a market economy. The population movement may originally have derived from southeastern Europe, but there does not appear to have been a repeated infusion of immigrants. Furthermore, after the initial colonization movement, there was "virtually no expansion of the settlement area" (Starling 1985:48).

Archaeologists have always treated Neolithic subsistence as if a single strategy applied to all local conditions. Neolithic farmers were generally assumed to have been shifting cultivators until very recently, when mounting evidence from Germany, the Netherlands, and Poland (Kruk 1980; Lüning 1982; Modderman 1970) began to suggest that many villages may have been permanently occupied. Apparently assuming a direct and positive correlation between cultivation strategies and village permanence, many archaeologists have reversed their opinions and flatly stated that Early Neolithic villagers practiced permanent field agriculture (see Dennell 1983). But village permanence and cultivation strategies are not necessarily correlated; moreover Central Europe is neither geographically nor climatically uniform. Its diverse environmental conditions suggest that not all regions would have been equally amenable to the same set of cultivation strategies. Instead, a variety of cultivation strategies probably existed. As discussed below, each strategy calls for a different mix of domestic and wild resources, and each would have had a different impact on the subsistence strategies of indigenous hunter-gatherers.

Shifting cultivation is a land-use strategy entailing the discontinuous cropping and vegetative regeneration of cultivated land during a fallow period. The fallow, an integral part of most cultivation systems, allows the soil to restore its fertility through the natural regeneration of organic content, microfauna, and mineral nutrients. Several types of fallow systems can be identified by the length of the fallow period and the vegetation that appears (Ruthenberg 1980:14–15):

Forest fallow allows the regeneration of the forest with a closed canopy. This is a long-term process and can take from one decade to several.

Bush fallow is a medium-term fallow, less than a decade in length. Shrubby vegetation develops but not a forest with a closed canopy.

Grass fallow is a short-term fallow with the regeneration of grasses but no woody vegetation.

Ley fallow is a special-purpose fallow in which grass or shrubby land is used for grazing. It can be short-, medium-, or long-term.

Each type of fallow results in different levels of fertility being returned to the soil. Nutrients taken up by and stored in the fallow vegetation are returned to the soil through the decomposition of the organic material and the development of humus. Woody plants are slower to grow, although they are larger and accumulate more nutrients than do grasses. Not surprisingly, forest fallows result in the greatest restoration of organic matter and nutrients to the soil, although forests take considerably longer to develop. As the closed forest returns, resource diversity first increases and then decreases. Prehistorically, forest margins around the open fields would have provided wild berries, hazel nuts, small game, deer, and boar; grass and ley fallows would have provided grazing for deer, while bush fallows would have supported deer, boar, rabbits, squirrels, and other small game, as well as berries and nuts. The regeneration of a closed-canopy forest results in less light reaching the forest floor; consequently, shrubby vegetation and floor growth decreases, and the abundance and diversity of resources declines.

Permanent cultivation differs from shifting cultivation in two ways. First, permanent cultivation entails the use of fixed plots for decades, with a regular rotation of crops grown on them. Fallows in such systems are generally short, involving only a few weeks or months between crops. In contrast, fallows in shifting cultivation are medium- to long-term, with a regeneration of bushlands,the mature forest, or a closed forest. Second, since the fallow in permanent cultivation is brief, vegetative regeneration cannot take place. Fertilization is generally required to maintain soil fertility, and the forest must be prevented from encroaching on the fields. Wild resources would be available along the forest margins, but abandoned fields in varying stages of long-term vegetative regeneration would not be available as a source of wild plants and game. In contrast, shifting cultivation results in patches in various stages of long-term vegetative regeneration with the attendant animal communities.

Permanent and shifting cultivation strategies are not mutually exclusive. Both can be combined with other subsistence strategies, and just as there are few, if any, pure hunter-gatherers, there are few pure cultivators. Three broad classes of cultivation systems can be identified (see Conklin 1954).

Integral shifting cultivators derive their primary subsistence from shifting cultivation, with secondary dependence on livestock, and/or wild foodstuffs (Conklin 1954). Cultivators who grow crops for exchange with other groups would also be included in this category. The ethnographic literature abounds with examples of integral shifting cultivators, including the Chembu (Brookfield and Brown 1963), the Hanunoo (Conklin 1957), the Iban (Freeman 1970), the Etero (Kelly 1977), and the Huron (Trigger 1976).

Integral agriculturalists are farming populations that subsist on foods obtained through fixed field agriculture.

Mixed cultivators derive their subsistence equally from shifting cultivation and stockbreeding, permanent field agriculture, and/or hunting-gathering-fishing. Ethnographic examples include the Nuer (Evans-Pritchard 1940), the Lamet (Izikowitz 1979), the Kofur (Netting 1968), and the Gwembe Tonga (Scudder 1962).

Partial cultivators place primary emphasis on hunting-gathering-fishing or stockbreeding, with only casual reliance on shifting cultivation. The best examples are hunter-gatherers such as the Menominee of Wisconsin (Hoffman 1896), the Agta in the Philippines (Peterson 1978a,b), the Maku of Brazil (Milton 1984), and the Cahuilla of Southern California (Lawton and Bean 1968).

Integral, mixed, and partial cultivation systems place different emphases on the importance of crops obtained from shifting cultivation. Although it is tempting to view Neolithic farming systems as homogeneous throughout Europe, it is unlikely that shifting cultivation was significant to the same degree throughout Central Europe in the Early Neolithic. In some areas, it may have been possible to have had systems with primary subsistence on permanent field agriculture; in others primary subsistence may have been on hunting-gathering-fishing rather than on shifting cultivation. In still others, mixed systems may have prevailed, with primary subsistence derived equally from several subsistence activities, including shifting cultivation. Finally, integral shifting cultivation systems may have been prevalent in other regions.

The most persistent problem with the term "shifting cultivation" lies with the widespread notion that a village must be relocated every few years. This is simply not true. The term "shifting" refers to plots of cultivated land, not to village permanence. Neolithic shifting cultivation in Central Europe has been compared to tropical swidden systems. In the tropics soils are thin, nutrients are leached away quickly from the root zone, and soil exhaustion often occurs. When cultivable soil near a village becomes exhausted, the village usually relocates to more fertile soils. Outside of regions with extremely thin soils, soil exhaustion is generally not a primary factor in village relocation. Infestation by vermin, threat of warfare, the death of a relative, persistent disease or sickness, repeated crop failures due to pests, and the inability of the villagers to resolve internal disputes have all been given as justifications for village relocation (Schlippe 1956:192–193; Smole 1976:86; Titiev 1944).

At least four different classes of residential mobility can be suggested for Neolithic populations:

Colonizing. Settlements are established, and the area around the village is cultivated for a given period of time. Rather than clear secondary or successional vegetation and remain in the same location, the group moves on.

Cyclical. Two or more settlements are used in rotation, with each village being occupied and vacated for specific periods of time.

Periodic. A village is occupied for an unspecified length of time, then abandoned, with no intent of reoccupation.

Permanent. A village is occupied for several centuries. The area around the village is systematically cultivated and fallowed.

Of the four, the colonizing strategy may have had the least effect on local hunter-gatherer populations. Colonizing swiddeners inhabit an area for a relatively brief time before moving on. Abandoned village sites would have been organically rich from accumulated habitation debris, and a luxuriant growth of vegetation would have developed and attracted game. In addition, escaped cultigens might have appeared for a while in old fields and gardens. Abandoned settlements could have provided a wealth of resources until the mature forest returned.

Cultivators with a periodic or a cyclical residential strategy would have had an even greater impact on hunter-gatherer subsistence strategies. Abandoned villages in various stages of forest regeneration would have contributed to the mosaic of forest resources. Of the two strategies, periodic village relocation probably would have provided the greatest degree of diversity. Moreover, since villagers would have planned on returning to previous settlement sites, they could have maintained vested rights to the resources at the unoccupied villages. Permanent villages would have had the most intensive effect on the forest, for the effects of the village would have been restricted to a limited area but over a long period. If the villagers practiced permanent field cultivation, long-term fallows would not occur. Unless the villagers practiced shifting cultivation, permanent villages would not necessarily have contributed to an increased mosaic of wild resources. As far as exploiting wild resources is concerned, hunter-gatherers would not benefit as greatly from a permanent village residential strategy as they would from any of the other three. In all the strategies, the presence of the villages would provide a predictable concentration of domestic resources.

Each of the four residential mobility strategies could be incorporated into an integral, mixed, or partial system of shifting cultivation. Thus a minimum of twelve potential subsistence-settlement strategies can be identified for the Early Neolithic period. Furthermore, the logistical mobility of the cultivators would also vary depending on the availability of wild resources and raw materials near a village. If dimensions of logistical mobility are added to the twelve basic subsistence-settlement strategies, a highly diverse set of Neolithic settlement systems must be expected.

Populations of indigenous hunter-gatherers could be articulated with the farmers in a number of ways. Fallow and abandoned plots would provide rich environments for wild resources. Moreover, the ripening crops themselves could attract some wild game, thereby increasing the density of some species near the villages. A complementary exchange of goods would allow each population to exploit a limited set of resources. The exchange would also make it possible for both populations to specialize and thereby increase their respective exploitive efficiencies while at the same time allowing both to enjoy a diverse resource base. The degree to which foraging populations would be affected by the farmers would be a function of the density of the cultivator populations and their residential mobility.

A strategy of either colonizing or permanent villages would not enhance the density and diversity of wild resources as much as would cyclical or periodic relocation strategies. However, a high density of Neolithic cultivators would probably severely affect indigenous hunter-gatherer populations and could ultimately result in their adoption of cultivation. At low densities, the two populations could not only co-exist; in all likelihood each would benefit from a cooperative interaction allowing both to specialize in a specific exploitive strategy. Conditions under which a cooperative interaction would develop are considered in the next chapter.

Summary and Conclusions

This chapter examined aspects of tribal and band level organization that were critical to the successful adaptation of a food-producing economy. Sedentism was not viewed as a critical prerequisite. Instead, mechanisms to prevent the over-exploitation of resources, to ensure for the production, storage, and allocation of crops, and to preserve seed grain for the next crop were viewed as being essential for food producers. Options regarding the residential mobility of shifting cultivation were reviewed, and a minimum of twelve possible land use strategies were identified for Neolithic farmers. The articulation of hunter-gatherers in any of these strategies would be highly variable.

3

Population Interaction

The first chapter concluded that assumptions underlying arguments for Neolithic expansion with either expulsion or acculturation of Mesolithic hunter-gatherers are untenable, and the second chapter identified organizational differences between small-scale populations with high and low residential mobility. This chapter develops a conceptual framework to investigate the interaction between foragers and farmers. Anthropology lacks a general theoretical construct for examining long-term interactions between populations with strikingly different economic strategies. But concepts of population interaction from evolutionary ecology have begun to prove useful in archaeological investigations. In particular, Spielmann (1982, 1986) explicitly developed an ecologically based model to explore the possibility of a mutualistic (cooperative) relationship between Plains bison hunters and Pueblo farmers in prehistoric North American.

Spielmann has shown that cooperative relations must be considered along with competition when examining the interaction of human populations. A general construct of population interaction should be effective in considering forager-farmer interaction during the so-called Mesolithic/Neolithic transition in Central Europe. Whereas Spielmann's North American work examines constraints affecting populations exploiting different environmental zones, forager-farmer relations in Central Europe involve populations inhabiting the same environment. This chapter reviews ecological theory and examines ethnographic examples of population interaction before developing a general construct of forager/farmer interaction for the mid-fifth millennium b.c. in the Alpine Foreland.

Throughout this discussion the foragers and the farmers will be treated as separate ecological populations. Since the concept of *population* is critical to investigating forager-farmer interaction, the term must be defined with particular reference to humans. Among ecologists there is much debate over its precise meaning. For most, it is an abstract concept, but ecologists generally subscribe to the definition that populations "are conspecific individuals occupying the same geographic region and who have a higher probability of ecologically and reproductively interacting with one another than with members of other populations" (Futuyma 1979:506; see also Pianka 1983:100). The key characteristic of a local population, then, is that it is a spatially constrained reproductive network.

Forager-farmer interaction occurs among members of the same species, *Homo sapiens*, and ethnographic data show that intermarriage between two groups may be prescribed by social rules. Strictly speaking, if a population is defined geographically as a reproductive network, then intermarrying foragers and farmers might be viewed as one population. However, their social and economic organizations are different. If the dimension of socio-economic organization is added to the definition, then distinct human populations can be defined *ad hoc* as clusters of individuals occupying the same geographic space, sharing a common socio-economic organization, and interacting socially and economically with one another more frequently than with members of differently organized groups.

The ecological concept of communities (defined as interacting populations [see Ricklefs 1979:669 ff.]) allows for the amalgamation of populations of foragers and farmers into a larger, more complex socio-economic unit. Throughout this book, the foragers and farmers are treated as separate ecological populations that form an ecological community if and when they interact.

Theoretical Framework

Ecologists have long recognized that interactions between two populations may be antagonistic, cooperative, or neutral. Over the past century ecologists have concentrated on investigating two types of antagonistic relations: competition and predation. They have paid little attention to cooperative interactions, preferring instead to treat them as "interesting but eccentric exceptions to the general rule" (Risch and Boucher 1976:8). Neutral relations have been dismissed entirely as being trivial and therefore uninteresting for evolutionary ecologists (Odum 1979; Pianka 1978). A growing number of researchers are beginning to examine the significance of cooperative interactions for the evolution, structure, and function of ecological communities (Addicott 1979; Axelrod and Hamilton 1981; Boucher 1979; Boucher *et al.* 1982; Dean 1983; Keeler 1981; Thompson 1982; Vandermeer 1980; Vandermeer and Boucher 1978; Wolin and Lawlor 1984). Their articles suggest that cooperation is probably as important as antagonism in structuring population interactions.

The construct developed here identifies critical variables shaping· cooperative or antagonistic relations. In particular, mutualism and competition will be examined. Predation—defined as the killing and consuming of one population (the prey) by another (the predator) (Pianka 1983:184)—between the two human populations is not included in the discussion. However, this does not eliminate the possibility that the foragers raided villages. If the foragers seized crops or livestock, the farmers and the foragers would not be in a predator-prey; instead, the two human populations would directly and very intensely compete for the domestic resources. Ecological investigations of neutral relations are so rare that it is impractical at this time to include neutrality in a general construct of prehistoric population interaction. This leaves competition and cooperation, the two ends of the interaction spectrum.

Competition

Competition occurs when two or more populations use a limited resource. Both participants are adversely affected by the relationship (Odum 1979), and throughout the interaction the number of both competitors may be reduced (Ricklefs 1979:238). Ultimately, one population will be more effective than the other, and it will probably survive while the other might not. Since competition depresses population levels and may even threaten the continued survival of a population, several mechanisms have been developed to reduce competition or to avoid it entirely. Dietary specialization is perhaps the most obvious (see Schoener 1977). It allows two populations to live within the same territory and exploit different resources. Contact and conflict are thereby avoided. The development of territoriality also allows resources to be partitioned and access restricted; moreover, territoriality permits a population to defend the general area in which resources occur rather than to defend each resource individually (see Ricklefs 1979:248–256).

Ecologists distinguish between direct and indirect competition. As the name implies, the former involves direct confrontation between populations using the same set of resources. It is a strong form of competition, which generally entails aggression between competitors and usually results in one population being prevented from using a set of resources (Ricklefs 1979:579). Direct competition requires expending energy in protecting resources, and injury or death can occur (Pianka 1978). Case and Gilpin (1974) note that direct competitors tend to be resource specialists, subject to stochastic factors affecting resource availability and that territory size fluctuates in response to population density. As population density increases, territorial infringements occur more often; consequently, the cost of maintaining the territory and the risk of injury increase. When costs of patrolling and defending territorial boundaries exceed the benefits derived from the resources, territory sizes may be restricted, and the population will have to intensify its resource exploitation (Case and Gilpin 1974).

Indirect competition occurs when one population uses or alters a resource thereby reducing its availability for another (Pianka 1983:185). The two populations do not come into contact; in fact, they may not even be aware of one another. Case and Gilpin (1974) suggest that indirect competitors (1) are generalists who exploit a variety of resources within their territory; (2) establish their territories in areas where the resources are the richest; (3) suffer little niche contraction in presence of direct competitors; and (4) tend to be sedentary. Indirect competition may be subtle; it does not entail direct confrontation, and is particularly effective when one population cannot out-compete another (Case and Gilpin 1974).

Dietary specialization and territoriality were introduced earlier as mechanisms by which populations could avoid or reduce harmful interactions, but neither eliminates them. Dietary specialists may engage in indirect competition if one population alters the resource base as a consequence of its subsistence activities. Territorial populations may encounter direct competition, if and when territorial infringements occur.

Both direct and indirect competition have been posited for the relationship between Mesolithic and Neolithic populations. On the one hand, it has been assumed that the foragers and farmers competed directly for territory within which to pursue their respective subsistence strategies. On the other, it has been asserted that the farmers competed indirectly with the foragers through forest clearance, which removed the game and plant resources the foragers relied on. It might be argued that because the farmers produced their own foods, they had a competitive advantage over the foragers, which would ultimately allow them to drive out the foragers. However, competitive exclusion occurs mainly in a saturated environment (Pianka 1983:193). As discussed in the first chapter, population growth in Central Europe was extremely low during the Early Neolithic. Similarly, Mesolithic population densities were quite low (see Jochim 1976). It is unlikely that during the Early Neolithic period the environment was saturated to the extent that competitive exclusion occurred. Furthermore, the archaeological record suggests long-term overlap between the two populations.

If both direct and indirect competition occurred, the relationship between the foragers and farmers must have been acrimonious indeed. From an ecological perspective, it would have been imperative for both populations to reduce competition lest each be adversely affected to the point that continued existence was jeopardized. Territoriality certainly could have been practiced, and archaeologists have noted a complementary distribution of Mesolithic and Neolithic sites (Tringham 1971).

Since the farmers settled along rivers, territoriality would have effectively removed riverine resources from the seasonal round of the foragers, and this would have required the realignment of forager subsistence strategies. Dietary specialization could have also been possible because the foragers depended on wild resources, while the farmers relied on domestic resources. But in producing their foodstuffs, the farmers may have restructured the resource configuration of the deciduous forest. This in turn could have reduced the forager resource base; but as discussed in the first chapter, it could also have enhanced the environment. In particular, the livestock and crops could have offered attractive resources and might have been subject to predation by the foragers. This would have then brought the foragers and farmers into direct competition.

Even if Mesolithic and Neolithic populations attempted to mitigate competition thorough dietary specialization or territoriality, competition could persist. Cooperation does not seem to be an obvious solution to competition; however, cooperative relationships appear to develop out of antagonistic ones (Thompson 1982), and cooperation offers a reasonable alternative to competition.

Mutualism

Mutualism occurs when two populations exchange goods or services to cooperatively exploit a range of resources. In contrast to competition, which is detrimental to the interacting populations, a mutualistic relationship is beneficial to both (Odum 1979). Mutualism encompasses a variety of relationships. Symbiosis, commensalism, cooperation, and proto-cooperation are often used, sometimes synonymously, to refer to mutualistic relationships (see Addicott 1984; Boucher *et al.* 1982). The ultimate criterion for identifying a mutualistic interaction is simply that the relationship results in higher carrying capacities for the interacting populations. Throughout this discussion, mutualism refers to the range of cooperative relationships that run the gauntlet from intense, physiologically linked interactions to loose relationships in which there is no direct contact between the interacting populations.

Mutualistic interactions can be divided into broad classes depending, among other things, on the degree to which the component populations interact, whether the relationship is a prerequisite for the continued existence of either population, the periodicity of the interaction, and the extent of reciprocal specialization (see Addicott 1984). This investigation is concerned primarily with the first two.

Ecologists distinguish between direct and indirect interaction according to whether the two populations have face-to-face contact. Direct mutualists actually provide one another with food, shelter, or services; whereas indirect mutualists create conditions leading to the increased availability of food or shelter, but the populations do not come into contact (Boucher *et al.* 1982). Direct interactions are further subdivided into categories according to whether the component populations are physiologically linked. Physiologically linked populations are referred to as having a symbiotic mutualism. For decades this was thought to be the most prevalent form of mutualism. Examples of the best-known symbiotic mutualisms include the gut flora responsible for the ability of ruminants to metabolize cellulose, fat cell bacteria allowing cockroaches to digest virtually anything, and nitrogen-fixing bacteria on the roots of legumes (Trager 1970). Non-symbiotic mutualisms are interactions that are not predicated on a physiological link between the component populations. Examples of non-symbiotic relationships include seed and pollen dispersal by birds, insects, and mammals (Janson *et al.* 1981); protective groomers that inhibit parasitic infestation (Smith 1968); and flocking behavior in birds (Cody 1971). The distinction between symbiotic and non-symbiotic interactions is unnecessary for indirect mutualism. By definition such mutualists do not directly interact, so a symbiotic relationship is precluded.

Ecologists also distinguish between obligate and facultative interactions. An obligate relationship is one that is absolutely essential for the survival of a component population: without the interaction the population would cease to exist. A facultative relationship, on the other hand, is one in which the population could survive in its absence, albeit at a lower carrying capacity.

Investigations of mutualistic interactions have begun to increase, but the vast majority of published works continues to present case studies examining the mechanics of specific examples[1] rather than develop syntheses aimed at understanding general conditions favoring the development and maintenance of mutualistic interactions. Nonetheless, although much work remains to be done, constraints affecting the development and maintenance of mutualistic interactions are slowly beginning to be identified. The following discussion briefly examines how mutualisms arise and reviews aspects of mutualism that appear to be significant for examining cooperative relationships among human populations.

Even though a definitive answer to the problem of how mutualisms arise has yet to be offered, Thompson (1982) suggests that two general avenues may exist. On the one hand, he argues that a complex relationship exists between antagonism and cooperation and that a wide variety of mutualisms have their origins in antagonistic interactions (but this is not to say that all mutualisms derive from antagonistic relations). The key to determining why some antagonistic relations continue to escalate and why some become cooperative is a relatively simple matter: "If it

---

[1]Boucher *et al.* (1982) provide an extensive bibliography of published studies through 1981.

is unlikely that individuals can avoid a specific antagonistic interaction, then selection will favor individuals that have traits causing the interaction to have at least less of a negative effect on them" (Thompson 1982:61).

On the other hand, Thompson argues that physical stress, such as is found in nutrient-poor habitats, may be the second major stimulus to establishing a mutualistic relationship (1982:88). In such environments, "small inputs by a mutualist can potentially have major effects on fitness" (1982:79). Unfortunately, other than reiterating that mutualism contributes to higher fitness levels, Thompson does not offer a mechanism for the development of mutualisms in nutrient-poor environments.

Non-symbiotic facultative mutualisms seem to be formed and dissolved fairly easily. Keeler (1981) notes that the development of a mutualistic relationship is favored when: (1) the majority of individuals establish effective interactions, (2) a pool of potential co-mutualists is both present and willing to participate; (3) participation entails low cost, (4) derived benefits are high, and (5) alternatives are ineffective. Thompson echos Keeler's first point, but he also adds that social behavior is critical for any mutualistic interaction, for one (if not both) populations may have to modify their behavior to participate in the relationship. Axelrod and Hamilton (1981) concur on Keeler's fifth point but go on to note that cooperation develops when the probability of repeated interactions is high. The points made by Thompson and by Axelrod and Hamilton are of particular importance, for a population may prefer to avoid contact by modifying its behavior and pursuing a feasible alternative, rather than to establish a mutualistic interaction. Conversely, it may choose to alter its behavior in order to enter into a cooperative relationship. Thus the ability to modify behavior may be a critical factor in whether a facultative interaction is actually formed.

Factors affecting the development of obligatory non-symbiotic mutualisms have not been firmly established. Roughgarden (1975) suggests that for obligatory symbiotic relationships to develop, there must be (1) significant fitness gains; (2) a high frequency of interaction; and (3) prospects of a long-term interaction after the relationship has been established. These vary little from Keeler's conditions favoring the establishment of non-symbiotic, facultative interactions and also include Axelrod and Hamilton's prerequisite. A critical factor in entering into either kind of interaction is that the benefit accrued must offset the requisite costs and behavioral modifications. As discussed below, the consequences of participating as an obligate partner appear to be potentially greater than those for a facultative one.

The most troublesome aspect of mutualistic relationships is that they may be asymmetrical. Co-mutualists need not depend on the interaction to the same degree: the participation of one population may be facultative and the other may be obligate (Dean 1983; Vandermeer and Boucher 1978). Mutualistic interactions can be quite stable, but the potential asymmetry of the relationship affects the stability and robustness of the interaction, as well as the potential survival of each population. If both populations are facultative mutualists, each can survive independently. But when the relationship is obligatory for at least one of the interacting populations, there is a threshold population level below which the interaction cannot be maintained, and the obligate mutualist may become extinct (Dean 1983; Vandermeer and Boucher 1978). If both participants are obligate, then each must at least maintain its threshold level: if *either* falls below its critical level, *both* may become extinct (Dean 1983; Vandermeer and Boucher 1978). The persistence of a symmetrical, obligate mutualism thus depends on the well-being of both populations.

Obligate mutualisms tend to develop in two types of environments. On the one hand, they occur in regions with relatively stable environments, such as the tropics (May 1981:95). This is understandable in light of the degree to which environmental perturbations are transmitted between obligate partners, with adverse conditions affecting one also jeopardizing the existence of the other. On the other hand, obligate mutualisms also occur in extreme and marginal environments (Lewis 1973). This is counter-intuitive since each population is susceptible to factors affecting their partner; however, in marginal environments, each mutualist can specialize in exploiting a limited set of resources that may demand large investments in exploitation strategies. An obligate interaction gives each population access to a network including both sets of resources, while simultaneously allowing both to limit their investment in skills, energy, and equipment in exploiting only one set of resources. Thus, an obligate interaction in a marginal environment may actually allow both populations to survive (Lewis 1973). Conditions favoring the establishment of a symmetrical, obligate mutualism must be extreme and the alternatives so restricted that each population is willing to cede its independence and rely on the other to maintain at least its threshold population and to fulfill its obligations, without which both would cease to exist.

Thus far, discussion has centered on characteristics favoring the development and maintenance of obligate interactions. It is clear from Keeler's points that facultative mutualisms are likely to develop when interaction with a potential pool of co-mutualists occurs regularly, the benefits of participating remain high relative to the costs incurred, and viable alternatives are ineffective. Conversely, when benefits decrease or a more viable alternative appears, the association can be disbanded. Interactions between facultative mutualists range from casual to intense, depending on the reliability with which each population produces the expected benefits, and on the role of the shared resource in each population's subsistence strategy. In addition, interaction can be intermittent, but nonetheless may recur at predictable intervals (Addicott 1984:448). Objectively determining the "costs" and "benefits" of participating in a mutualistic interaction can be extremely difficult, for the benefits derived often cannot be directly compared with the cost of producing the goods or performing the services. Furthermore, it may be difficult objectively to estimate the relative "value" of reciprocal goods and services, for the worth of the benefit is defined by the recipient population.

Ecological case studies illustrate variables that appear to be critical in the functioning of mutualistic interactions. As discussed earlier, direct mutualism entails the exchange of benefits. These benefits can be classed into three categories: nutrition, protection, and transportation. Nutritional benefits are a component of most exchanges, and they are an important element of mutualistic interactions. Protection appears to be the most common benefit. For example: ants protect aphids and plants from predators in exchange for honeydew and nectar (Addicott 1979; Keeler 1981); ants obtain honeydew from butterfly larvae in exchange for protecting them from parasitic wasps (Pierce and Mead 1981); sea-anemones protect small fish, which consume waste from the sea-anemones (Trager 1970); and parasitic cowbird chicks groom host nestlings, thereby protecting them from parasites while obtaining food from the host parents (Smith 1968). Examples of transportation include birds and insects that carry pollen from one flower to another (Dressler 1982; Howe and Smallwood 1982; Regal 1982), thereby ensuring the plants's reproduction while obtaining nectar and pollen as food; squirrels and other mammals that cache acorns as food stores, thus placing the oak's seeds in environments favoring germination (Boucher 1979, 1981); and ruminants that ingest seeds while grazing and transport them to new locations.

These examples illustrate several points. First, the exchange is complementary: food is commonly exchanged for resources or services but rarely for other food. This is important, as it contributes to decreasing the potential for competition. In examining the stability of mutualistic relations, Addicott (1978, 1979, 1981), Boucher (1979), and Wolin and Lawlor (1984) have concluded that many mutualistic interactions are density dependent. At low population levels the relationship may be mutualistic while at high population levels it may be antagonistic. Thus, the relationship shifts from cooperative to antagonistic depending on environmental conditions (Boucher *et al.* 1982).

Second, each population provides a benefit that is either an item produced in excess of the producer's normal uses, or it is a by-product of the population's food procurement or other activity. In either case, benefits derived from the service outweigh the marginal cost of producing the nutritional resource. Third, the interaction occurs either between two mobile populations in a specific location or between a sedentary and a mobile population. The benefit of the interaction must be high enough for both mobile populations to modify their behavior to repeatedly meet at the location; the benefit accruing to a mobile population must outweigh the cost of changing its locational behavior to accommodate the sedentary population; and the service provided to a sedentary population must exceed the cost of producing the resource to attract a mobile population.

So far, the discussion has examined theoretical aspects of mutualistic interaction derived from the observation of insects, birds, and fishes. All of this must be applied to the problem of forager-farmer interaction during the fifth millennium b.c. Two factors are critical. First, this study concerns interactions of human populations, between whom there is no physiological link; therefore, we are concerned with a non-symbiotic relationship. Non-symbiotic interactions tend to be facultative, rather than obligatory (Boucher *et al.* 1982). If the foragers and farmers entered into a mutualistic relationship, it would most likely have been facultative. Both populations would have been able to retain their independence, and neither would have been dependent on the other for continued survival.

Second, facultative interactions seem to operate best under conditions of low population density (Addicott 1979). As discussed earlier, population densities of both the foragers and the farmers appear to have been low in Central Europe during the mid-fifth millennium b.c. This does not, however, ensure that a mutualistic interaction in fact developed. For a cooperative relationship to develop it would have been necessary for the populations to have interacted sufficiently with one another to realize that (a) each offered the other goods or services the other was unable or unwilling to produce; (b) viable alternatives were not feasible; and (c) the benefits derived from the relationship warranted any requisite behavioral modifications.

Milisauskas (1978) has already suggested that the foragers and farmers may have exchanged cereals for wild resources, particularly game. Such a direct interaction may have been possible; alternatively, other items of exchange may have been used. Indirect mutualism may also have occurred. As discussed in the second chapter, if the horticulturalists practiced a long-term fallow, then the forest near their villages would include a patchwork of plots in various stages of succession with their attendant animal populations. A third possibility is that a relationship may have cycled between antagonism and cooperation. The ethnographic record provides one avenue of determining whether human populations actually enter into mutualistic relationships, the nature and intensity of the interactions, and the classes of goods and services that might be important to either direct or indirect mutualists.

Conditions Favoring Forager-Farmer Mutualism

As discussed above, mutualistic interactions develop because two populations provide each other with resources, services, or benefits allowing each to increase its carrying capacity. A number of resources and services may be exchanged; however, the exchange of subsistence resources is the one most likely to have occurred during the Mesolithic/Neolithic transition, and it may be the easiest to identify archaeologically. Four ethnographic examples of a mutualistic relationship are examined. Before considering the ethnographic examples, it is necessary to examine two aspects of human nutrition that are important in investigating mutualistic interactions.

First, proteins provide amino acids needed to build, repair and/or replace body tissues, while carbohydrates and fats provide energy in the diet. The body's basic energy requirements must be met before repair and maintenance activities are carried out. If sufficient energy is not available through carbohydrates and fats, then dietary—and when necessary body—proteins will be used for fulfilling basic energy needs rather than for body growth and maintenance. Second, food has a thermogenic effect in that digestion causes the basic metabolic rate to increase; proteins raise it the most, fats second, and carbohydrates the least (Davidson *et al.* 1979:19). Individuals subsisting on diets consisting primarily of protein have a higher basic metabolic rate than individuals whose diets include moderate quantities of carbohydrates and fats. Consequently, individuals with a protein-rich but fat- and carbohydrate-poor diet have higher minimum caloric requirements than individuals with moderate quantities of fats, proteins, and carbohydrates.

Speth and Spielmann (1983) note that hunter-gatherer populations are particularly susceptible to dietary stress during seasons in which game animals are depleted of their fat reserves. Furthermore, if shortages of carbohydrate-rich plant foods regularly occur at the same time, the population would have to develop mechanisms for either storing or exchanging foods rich in fats and carbohydrates. Conversely, populations subsisting primarily on carbohydrates and fats would eventually suffer from protein deficiencies. Populations with insufficient proteins must obtain either animal protein or protein-rich plant foods with amino acid patterns complementary to the plant foods they already consume.

The most common mutualistic interaction to expect would be the exchange of protein resources for fats and/or carbohydrates, such as Spielmann (1982, 1986) has found in her research. This type of interaction could take any of several forms. The simplest would be for two populations to respectively specialize in the production procurement of proteins, fats, or carbohydrates and then exchange with another population for the other.

Specialization-and-exchange is likely to develop when each extractive technology requires special skills, training, or intensive labor or where each resource can be obtained only at discrete and distant locations. Such an interaction is most likely to occur between populations inhabiting environmental zones with sharply different resources (see Spielmann 1982, 1986). Specialization would allow each population to exploit its region more intensively, while exchange would ensure that both populations obtained the resources they lacked. Specialization-and-exchange also provides one way of mitigating competition between two populations in the same environment. If two populations were generalists, both would exploit the same set of resources. This would result in competition for resources if and when population densities increased. With specialized procurement strategies, both populations could intensify their resource extraction without a concomitant increase in competitive interactions.

Another likely form of mutualism is that one population may have a subsistence economy with a balance of proteins, carbohydrates, and fats, while a second population may be deficient in any of these. From a nutritional perspective, the former would not need to enter into a mutualistic relationship. However, they may lack critical non-food resources, have an insufficient labor force to perform production activities, be vulnerable to predation from animals, or subject to raids from a neighboring population. Here again, mutualism offers an alternative to antagonistic interaction. Cooperation would allow the latter population to provide a needed service in exchange for the resource it might otherwise take by force. Finally, social or religious proscriptions may prevent one population from performing a given task. Once again, a cooperative relationship would allow one population to provide a needed service to the second population.

As discussed above, mutualistic relations will develop only when there is a likelihood of repeated interaction, when the benefits derived offset the costs incurred, and when satisfactory alternatives are not available. Anthropologists have long recognized cooperation between social groups; however, it is only recently that some of these interactions have been identified as being mutualistic. Spielmann (1986: Table 3) lists 25 ethnographic examples of mutualistic interaction between hunter-gatherers and other hunter-gatherers, hunter-gatherers and food producers, or food producers and other food producers in arid, temperate, and tropical environments. Prehistoric mutualistic relations have also been identified in the American Southwest (Kelley 1986; Spielmann 1982), and mutualism was undoubtedly an important aspect of the Huron trade network in the Great Lakes region (see Heidenreich 1971, 1978:383–385).

Evidence for mutualism between temperate forest wheat farmers and foragers is sorely lacking in the ethnographic record.

A review of four ethnographic examples of forager-farmer interactions allows the critical parameters affecting the general nature and intensity of mutualistic relationships to be identified.

Ethnographic Examples

Mbuti-Bantu Interaction

The Mbuti are hunter-gatherers and the Bantu are horticulturalists who inhabit the equatorial rain forest of Zaïre (Harako 1976; Hart 1979; Hart and Hart 1986; Tanno 1976). They enjoy a relationship in which the Mbuti hunt and provide animal protein in exchange for carbohydrates from the Bantu. Game is an important component of each population's diet, but it has a low fat content and must be complemented by plant foods (Hart and Hart 1986). A small number of wild plants in the closed evergreen forest are carbohydrate- or oil-rich; but these species are uncommon, and none of the calorically important species are available in the primary forest (Hart and Hart 1986). The important species are abundant only in the savanna ecotone, gallery forests, and secondary forests succeeding abandoned swiddens. The Bantu provide a direct source of carbohydrates through planting and harvesting yams. Their cycle of forest clearance and abandonment also provides an indirect source of carbohydrates, for feral cultigens and economically useful pioneer plants thrive in abandoned swiddens (Hart and Hart 1986).

The relationship between the Bantu and Mbuti is asymmetrical. The Bantu could hunt themselves, although hunting would take time away from their horticultural activities; therefore, their participation is facultative. Whereas the Bantu could survive without the Mbuti, the converse is not true. The Mbuti can survive in the equatorial rain forest only because the Bantu provide carbohydrates directly in exchange for meat and indirectly as by-products of their horticultural activities (Hart and Hart 1986). Thus, the Mbuti participate as obligate mutualists in the relationship. Without the farmers, the Mbuti would not be able to inhabit the forest interior.

Bihor-Horticulturalist Interaction

The Bihor are hunter-gatherers in India who participate directly in a market economy. Each band intensively exploits a single ecozone, and it cannot move across territorial boundaries to fill seasonal lows (Smiley 1981). They trade game and manufactured goods in the market place in exchange for rice, which comprises 80% of the Bihor diet. The meat and manufactured goods obtained from the Bihor free the villagers to devote their attention to production activities, but the Bihor may be providing a more significant service. Hindu villagers either do not eat any meat or at the very least have taboos against major meat sources, such as monkeys. The hunter-gatherers may protect the villagers' crops by removing predators, which the villagers are not allowed to kill themselves. In either case, the relationship is asymmetrical: the Bihor could not exist as hunter-gatherers without the carbohydrates supplied by the horticulturalists, while the horticulturalists probably could survive without the services and resources supplied by the Bihor. If this is true, Bihor are obligate mutualists and the villagers are facultative.

Maku-Tukanoan Interaction

The Maku, hunter-gatherers in the Black Water area of Brazil, trade game in exchange for carbohydrates from the Tukanoans, who are horticulturalists (Milton 1984). Like the African equatorial rain forest, the interior forest in the Black Water area does not provide abundant sources of wild carbohydrates. Carbohydrates from the Tukanoans provide 80% of the Maku diet, and the Maku would not be able to inhabit the forest without their relationship with the Tukanoans (Milton 1984). The Tukanoans do fish, and thus obtain some of their protein. However, fishing is time-consuming, and their relationship with the Maku allows them to devote more time to horticultural activities.

Agta-Lowlander Interaction

The Agta are hunter-gatherers in the Philippines who occasionally plant a small swidden, but prefer to exchange forest products and wild game for corn, rice, and western manufactured goods from neighboring horticulturalists in the Palanan lowlands (Griffin 1984; Peterson 1978). Unlike the previous examples, the interaction is symmetrically facultative. The Agta could broaden their subsistence strategy to include regular cultivation of forest plots; however, it would bring them into competition with the Palanan lowlanders for productive lands. The Lowlanders themselves could go into the forests to obtain the requisite forest products, although this in turn would intensify competition with the Agta.

Mutualism Between Human Populations

The above examples conform to the prerequisites for mutualistic interaction set forth by Keeler (1981) and Roughgarden (1975). Ethnographic instances of mutualistic relationships are all founded on a network of personal ties and include most members of the population in the interaction. For example, the Agta-Palanan lowlander and Mbuti-Bantu interactions, which entail long-term relationships between household of hunter-gatherers and cultivators, are intentionally maintained and may be sustained for generations. In the Agta-Palanan interaction this partnership constitutes a named, cultural category; it entails a commitment to regular exchanges, and allows for an extension of credit and other forms of mutual assistance (Peterson 1978b:80).

Among the Mbuti and Bantu, obligations are maintained by cross-cultural marriage and fictive kin ties. A child of a cross-cultural marriage has legally sanctioned rights within each respective social system. This incorporates both partners in a kinship network entailing reciprocal responsibilities and obligations, and trading is frequently conducted within the framework of kinship obligations (Hart 1979:71–76).

The Bihor example is somewhat different. Individuals take their game and manufactured goods to the market place, where it is then sold or traded. Stability in the relationship is provided by conducting the transaction in the same location under the auspices of the market system, rather than with the same individual. The market demand for Bihor produce is steady, and both parties are well aware of the benefits they can expect to derive from the interaction (Williams 1974).

In all four cases participation entails relatively low costs while yielding relatively high benefits for the parties involved. The Agta, Bihor, Mbuti, and Maku procure game or manufacture items at little cost beyond what they normally expend in such activities. Peterson (1978b) argues that a deer or wild pig yields more meat than an Agta group can eat; therefore, trade provides an efficient use of meat that would otherwise have to be abandoned. Conversely, it might be argued that the Palanan lowlanders expend less energy to grow surplus corn, yams, or rice for trading to the Agta than they would to produce fodder and manage herds large enough to provide an equivalent amount of domestic animal protein throughout the year. The interaction is probably more complicated (see Headland 1978). First, deer and wild pigs in the Philippines are small, so it is unlikely that an Agta band would have trouble finishing one of these animals (Parry, personal communication). The significant aspect of maintaining the exchange is that both groups provide items for which there is no alternative source: the Palanan lowlanders supply manufactured items in exchange for medicinal plants and rattan.

Similarly, the Bihor, Mbuti, and Maku supply their partners with meat at regular intervals and in relatively small, predictable quantities. In each instance the horticulturalists could hunt themselves or they could raise livestock; however, it is advantageous for the villagers allow the hunter-gatherers to bring meat to the village. In all cases, the component populations obtain a resource which they cannot or will not produce themselves.

The alternative, for each population to produce the resource itself, is ineffective. In order for the Agta, Maku, Bihor, or Mbuti to intensify their own procurement of carbohydrates, they would have to increase their own horticultural activities, for suitable wild sources of carbohydrates are not available in sufficient quantities in their respective forest habitats. This would result in an abdication of their freedom of mobility in order to tether themselves to specific resources. As discussed in Chapter 2, reducing residential mobility requires a suite of changes to allow for increased contact between individuals. Moreover, by increasing the intensity of their own horticultural activities, the hunter-gatherers would come into direct competition with the horticulturalists. Regardless of the outcome, intensive production of domestic carbohydrates would result in the end of a mobile hunter-gatherer social organization.

From the horticulturalists' perspective, procuring animal proteins would entail intensifying either stock breeding or hunting and fishing. It would also require an increased investment in tools and training needed for either, as well as the time needed to hunt or to care for the livestock. This in turn would reduce the time available for horticultural activities. Finally, such a broadening of the subsistence base would bring them into direct and indirect competition with the hunter-gatherers for both hunting territories and additional grazing land.

Overall, mutualistic interactions in the ethnographic literature meet the prerequisites set forth by Keeler (1981) and Roughgarden (1975). The majority of individuals establish effective interactions, participation entails low cost, derived benefits are high, and alternatives are both ineffective and undesirable. Ethnographic studies over the past two decades have also made it possible to identify critical factors affecting the stability of forager-farmer interactions. When both populations have access to a complete set of resources, the interaction is more likely to be symmetrically facultative. Asymmetrical mutualistic interactions tend to occur when one population lacks a critical element in its resource base. (In the majority of cases it has been carbohydrates.)

Although the ethnographic sample is small and was not statistically drawn, the high frequency of obligate mutualists contradicts the observation by Boucher *et al.* (1982) that nonsymbiotic mutualisms are most likely to be facultative. Then again, the obligate mutualists all inhabit forest interiors that offer few carbohydrate resources. It is precisely because of the carbohydrate-producing partner that the obligate foragers are able to exploit the forest. Some ethnographic examples thus support arguments by Thompson (1982) that mutualistic relationships may develop in nutrient-poor environments; furthermore, many environments may not have been open to exploitation until the appearance of cultivators, who could provide a reliable source of carbohydrates. As noted earlier, asymmetrical mutualisms are more unstable than are symmetrical interactions, and this has implications for forager-farmer interactions. The carbohydrate-rich facultative partner is far freer to determine the terms and conditions of the interaction than is the carbohydrate-poor, obligate mutualist.

Prehistoric Interaction

So far the discussion has concerned theoretical aspects of competition and mutualism. These must be applied to the specific problem of the relationship between Mesolithic foragers and Neolithic farmers in the Alpine Foreland. Issues to be resolved include determining the effects competitive or mutualistic interactions would have had on each population, whether each population could have provided resources or benefits for the other, and factors that would have affected the stability of the interaction.

Neolithic populations would have competed both directly and indirectly with indigenous Mesolithic populations. Indirect competition could have resulted from establishing villages, clearing forested areas or grasslands for cultivation, allowing livestock to graze in natural pastures and to browse in the forest, and harvesting natural meadows for winter fodder. These activities would have resulted in major modifications to the deciduous forest. However, they may not have had a deleterious effect on wildlife diversity and abundance. Abandoned fields as well as uninhabited villages and homesteads would have resulted in a mosaic of gaps in the deciduous forest. Forest gaps and margins contain a variety of shrubs, grasses, and herbs not found in the forest interior. This vegetation would have attracted game, particularly deer and wild pigs. Equally important, a variety of berries and other economically useful plants thrive along forest margins. Of these, hazel nuts would have provided a particularly significant resource for human populations.

Whether the activities of the farmers enhanced or degraded resource availability for hunter-gatherers would have depended on the number of villages, their size, and the frequency with which they were abandoned and reestablished. Low population densities, few villages, and infrequent village relocations would result in the addition of a relatively small number of openings in the forest, thereby increasing the net diversity of the forest without altering its general composition. High population densities, numerous villages, and frequent moves would have had a greater impact. At some point the forest would not have had sufficient time to regenerate, and the availability of game ultimately would have begun to decline.

Cereals, milk, and meat could have provided the bulk of a village's nutritional needs. The predictability and reliability of crop yields and meat production would have affected the degree to which Neolithic populations could subsist on domestic resources. Limited gathering would have added nuts, berries, and greens to the diet, but intense gathering, hunting, and fishing could have provided a significant proportion of the Neolithic diet. This would have brought Neolithic populations into direct competition with the Mesolithic foragers.

On the other hand, the crops and livestock themselves may have been attractive resources for Mesolithic hunting-gathering populations. A mixed oak forest itself offers a few easily stored, high-energy resources. The resources are limited primarily to hazel nuts (*Corylus avellana*) and acorns (*Quercus robur* and *Quercus petraea*), but there is no archaeological evidence, such as charred remains, to suggest acorns were exploited by either Mesolithic or Early Neolithic populations in Central Europe. Sweet grass (*Glyceria fluitans*) and cattail (*Typha latifolia*) form large stands along lake margins; both would have provided good sources of carbohydrates. Here again tools and equipment needed to process either resource have not yet been found at Mesolithic sites in the Alpine Foreland, so their exploitation cannot be confirmed. Finally, beech (*Fagus sylvatica*) and water chestnut (*Trapa natans*) slowly began to spread throughout the Alpine Foreland in the Early Neolithic, but there is little evidence to suggest the widespread use of beechnuts or water chestnuts before the Middle Neolithic.

Unlike prehistoric Eastern North America, where the offsetting mast cycles of several species of nut-bearing trees were exploited, archaeological evidence suggests only one species, hazel, was exploited in the Terminal Mesolithic and Early Neolithic of Central Europe. Hazel masts occur once every five or six years. The nuts may have been an important food, but probably would not have provided a predictable resource every year. Postglacial hunter-gatherer populations obviously survived for millennia in the Alpine Foreland, and they must have obtained wild carbohydrates. With the arrival of Early Neolithic farmers, however, carbohydrates could have been obtained more easily and more reliably from the cultivators than from wild sources. This would have allowed a higher density of foragers to inhabit the region.

Livestock may have also been an attractive resource for the hunter-gatherers. Hunting is difficult in the late winter because of melting snows and the lack of forest browse. Fats are particularly critical in the human diet at this time, but the meat of game is lean in late winter (Speth and Spielmann 1982). Neolithic livestock offered more than a ready source of protein. Cow, goat, and sheep milks are high in fats, and milk becomes available in the very late winter. Obtaining meat and/or milk from the farmers would have greatly enhanced the ability of the hunter-gatherers to bridge the late winter/early spring when critical resources were in low supply.

Antagonistic relations could develop between foragers and farmers over domestic resources. The foragers may have raided

Neolithic villages for either crops or livestock or both. Periodic raiding would have forced the farmers to incorporate wild foods as significant items in their diet. This would have brought the Neolithic populations into direct competition with Mesolithic foragers for forest resources. Moreover, raids would have stimulated the need for defensive protection. Open conflict would have resulted in a risk of injury for both populations.

Finally, acculturation would have put the Mesolithic foragers in direct competition with the Neolithic farmers. However, it is questionable whether foragers would have been willing to adopt a Neolithic lifestyle. The success of a mixed farming economy was predicated on a certain degree of environmental manipulation and transformation. Producing domestic foodstuffs, therefore, required a greater labor investment than did hunting and gathering. If foraging populations could obtain domestic resources at little cost and without greatly altering their foraging lifestyle, then it seems logical that they would have preferred to continue their Mesolithic existence.

A mutualistic interaction would have avoided open conflict by allowing the farmers to control the amount of domestic fats or carbohydrates given to the foragers. In return the hunter-gatherers could have provided three potential benefits: (1) protection from wild game (particularly wild pigs and deer) who feed on ripening grain; (2) provision of labor during planting or harvest seasons; or (3) transportation of forest products, raw materials, or information. It is possible to envision a non-symbiotic facultative interaction between the foragers and farmers. The farmers would need to increase food production above the village's needs. The foragers would need to modify their subsistence activities to accommodate labor requirements during planting and harvest seasons, but only minor alterations might be required in their hunting strategies to remove wild pigs and deer from crops or in their mobility strategies to transport information between villages.

Mutualistic interactions need not be balanced in terms of both populations participating to the same degree or receiving an equal benefit. Each population need only accrue benefits that offset *either* the costs incurred *or* the costs of pursuing alternative choices. The decision to continue or to disband the association would rest on each population's constant reassessment of the benefits received and the relative costs of alternative strategies.

Summary and Conclusions

Both competitive and mutualistic interactions can be suggested between immigrant farmers and indigenous foragers. Competitive interactions develop when two populations exploit the same set of resources, or when one population interferes with the resource base of a second population. Direct competition is likely to develop between territorial populations when one territory lacks critical resources, or when it becomes unable to support its population. Indirect competition develops when one population alters the resource base of another.

Mutualistic interaction allows territorial populations or dietary specialists to avoid competition by regulating their interaction. Cooperative relations develop when two populations provide one another with a resource or service that costs relatively little to produce but which the other population is either unwilling or unable to produce itself. Maintaining a cooperative interaction depends on each population's willingness to participate. Costs must remain low relative to the benefits derived, alternatives must be less attractive, and the interaction must be regular and predictable. Mutualistic interactions are dynamic, relating to predictable as well as stochastic fluctuations in the resource base. The production cycles of both partners need not be synchronous. One partner may produce its resource or service sporadically while the other steadily provides its benefit. An interaction can appear to be antagonistic in the short run and cooperative in the long run. Alternatively, it can fluctuate annually.

Forager/farmer interaction depends on a series of factors. The costs incurred by and benefits derived from an interaction can be determined only with reference to each population's resource configuration, technological base, and scheduling of subsistence activities. The next three chapters examine basic issues such as crop productivity, animal husbandry, and domestic resource fluctuations in the Neolithic diet, before examining factors shaping terminal Mesolithic subsistence, and the interaction itself.

4

Neolithic Subsistence I: Crops

The two previous chapters examined anthropological and ecological aspects of population interaction. In Chapter 2 integral, mixed, and partial cultivation subsistence strategies were introduced. Each places a different reliance on cultivated crops, and it is likely that all three existed in Neolithic Europe. Each would have had different implications for forager-farmer relationships. At one end of the scale, integral cultivators would require relatively large areas for crop land; thus the potential for indirect competition might be great. At the other end, partial cultivators would depend heavily on wild resources; therefore direct competition was likely to be greatest among partial cultivators and foragers. A third aspect, the effects of stockbreeding, must also be considered. Mixed and partial cultivators might have depended heavily on stockbreeding, and livestock grazing and browsing may have affected large areas of forest. This could lead to an increase in indirect competition.

The potential for direct and indirect competition may have been high, but the possibility of mutualistic interaction was also high. Integral cultivators may have encountered scheduling problems, and the foragers may have provided one means of alleviating these problems. Mixed and partial cultivators may have found resource partitioning with subsequent specialization and exchange to have been a reasonable alternative for avoiding direct and indirect competition.

To fully investigate forager-farmer relations, it is necessary to identify the constraints affecting integral, mixed, and partial cultivators; to evaluate the farmers' ability to produce sufficient surpluses that would allow them to enter into a cooperative relationship; and to determine the areas in which the farmers might compete directly with the foragers. The best way to resolve these questions is through a technical examination of Neolithic farming. This chapter, therefore, examines the ecology of crops and weeds commonly found at Early Neolithic sites, discusses potential prehistoric crop yields, and suggests a general model of cropping and planting strategies. Data presented in this chapter are used later to develop a dynamic model of optimal farming strategies.

Cereals

The life cycle of wheat can be divided into seven major phases (Gill and Vear 1980; Percival 1921)—germination, tillering, stem elongation, shooting, flowering, grain maturation, and ripening. The shoot usually appears above ground within two weeks of planting, and secondary shoots (called tillers) begin to form. Tillers develop for several weeks while the plants remain low to the ground, with the leaves telescoped inside one another. At a given point, leaf formation ceases and the ear develops within the protective cover of the leaf sheaths. This marks the end of leaf formation. Each stem then begins to elongate, carrying the ear upward and ultimately into the open air. Pollination and fertilization occur shortly after the ear appears, and the grain subsequently matures over a period of several weeks (Percival 1921). Once the grain has matured, it moves through several stages of ripeness within a period of 10–14 days (Scheffer 1972). Grain passes through a milk ripe stage; a soft dough stage; and a waxy ripe stage. In rapid succession first a full ripe, and then a dead ripe stage follow. Protein and mineral content are highest at the dead ripe stage (Scheffer 1972), but by then the ears are so brittle that they can shatter, and harvest losses can be high. The grain is best harvested at the full ripe stage.

Factors Affecting Yield

Climatic regime is the single most important factor in determining overall crop yields (Bourke 1984; Sakamoto 1981). Typically, cereal crops require a cool germination period; moderate to warm temperatures during the vegetative and maturation phases; and a warm ripening period (Gill and Vear 1980; Percival 1921; Zimmerman 1950). Isolated events—such as severe weather; insect, bird or mammal predation; and infestation by rusts, smuts, and blights—can result in low yields even in years that otherwise would have produced bumper crops. The most critical factors are considered below.

Sowing Time

Day length controls the development of the ear, so changing the time of sowing does not significantly affect the time of flowering and harvest (Gill and Vear 1980:54). The amount of tillering does depend on the length of time between sowing and the occurrence of inflorescence development. The longer the period, the more tillers, hence the greater the number of ears per plant. Cereals can be planted in either fall or spring, and they are commonly referred to as being "winter" or "spring" crops depending on the season in which the crop is sown.

A winter crop is sown in the fall and harvested some nine months later in the following summer. During the fall the plants develop a small rosette of roots and tillers before winter sets in. Snow then provides a blanket that protects the young plants from the extreme cold of winter. When the snow melts in the spring, winter crops continue developing their vegetative structures. Winter crops are open to a number of hazards. If the fall is too cold and wet—or alternatively too dry—germination will not occur, and the crop must be replanted in the spring. If the winter is mild and thaws frequent, frost heaving occurs; delicate root structures become exposed to the elements, as well as to bird or mouse predation, and the plants are not protected from blasts of frigid air. On the other hand, when a winter is too severe, the crop's cold tolerance may be exceeded, and the crop can be lost.

Spring crops are planted in the spring and harvested three to four months later. The plants germinate and develop a root system before beginning to tiller. This can take three to four weeks, and, as a result, spring crops can be up to 30% smaller than winter crops. Spring crops may not be subjected to the vagaries of winter weather, but they are vulnerable to hazards surrounding seed storage. Mouse predation can lead to serious seed grain loss, as can the development of bacterial growth in storage containers. A cold wet spring can delay germination to the extent that the vegetative phase can be shortened considerably. Finally, spring crops are not as cold-hardy as the winter crops, so they are more susceptible to damage resulting from late frosts.

Soil Moisture

Plants draw practically all their water and nutrients from the soil. When transpiration exceeds water uptake, the plant is stressed: its metabolism changes, and growth slows. Water stress is more harmful at some stages of plant development than at others. If stress occurs early during the vegetative phase, tillering and root establishment are checked; moreover, grain yield is not necessarily affected. When stress occurs during the development of the florets, ears may form with fewer florets, and the overall number of grains per ear can be reduced (Briggs 1978; Percival 1921). Stress during pollen production is more devastating. The formation of fertile pollen can be inhibited, and sterile ears can result.

Excessive soil moisture can be equally damaging (Palti 1981). When soils are saturated, waterlogging occurs. At best, root development and tillering are retarded. More often, waterlogging prevents the rootlets from absorbing oxygen, and the plant literally suffocates (Bourke 1984; Briggs 1978; Percival 1921). As with drought stress, waterlogging affects yields at some phases more seriously than at others. Even if the new crop is unaffected by excess soil moisture, yields will be low simply because of the shortened tillering period. Damage is most severe during the heading and flowering stage, for fertilization is inhibited (Briggs 1978; Gill and Vear 1980) and a lower yield results. Finally, when the soil is waterlogged, field humidity is high and an environment develops in which a variety of fungoid diseases thrive.

Soil Fertility

Wheat is nitrogen-poor and requires large quantities of nitrogen to maintain vigorous growth (Scheffer 1972). Normally, when a plant dies it decomposes and nutrients held up in its tissues are returned to the soil. In the case of domestic cereals, the ears and straw are usually taken from the field. The nutrients bound up in the plants are thus removed from the cycle. When high-yield cereals are grown repeatedly on the same plot, soil nutrients may be exhausted, with nitrogen depletion occurring at a faster rate. Since nitrogen is responsible for the protein quality of the grain, the lack of nitrogen in the soil can affect the nutritional quality of the grain, and result in decreased crop yields.

A long-standing assumption held by prehistorians has been that soil exhaustion occurred readily throughout the Neolithic

Table 2

Nitrogen Exchange Without Manuring

Wheat Yield (kg)	Nitrogen Removed by Crop (kg)	Nitrogen Replaced by		
		Dust, Rain Birds (kg)	Seed Grain (kg)	Leguminous Weeds (kg)
1000	20	8–12	4	2–10

Source: Loomis 1978:480

(Childe 1929; Clark 1952; Sangmeister 1951). The assumption appears to be unwarranted. Until this century, wheat yields fluctuated near 1000 kg/ha. When annual wheat yields fluctuate at roughly 1000 kg/ha, nitrogen removed by the wheat crop is replaced by dust, rain, birds, and leguminous weeds (Loomis 1978), and by the seed of the cereal crop itself (see Table 2). Manuring is not necessary to maintain soil fertility, and stable crop yields can be expected almost indefinitely (Loomis 1978). The arguments are borne out by long-term crop studies at Rothamstead, England (Cooke 1976; Hall 1917; Johnston and Mattingly 1976). Wheat yields initially decline, and subsequently stabilize at about 1000 kg/ha. Fluctuation thereafter appears to arise more from climatic variations than from soil fertility.

Rusts, Smuts, and Blights

Diseases such as rusts, smuts, and blights include a plethora of fungal spores that attack the crop (Martin and Salmon 1953). Once the disease is established on a plant, it uses nutrients and water that otherwise would have gone toward grain development. When parasitic diseases become established, crop losses can be as high as 80–90%. The occurrence of rusts, blights, and smuts is a constant threat, but infection is local, sporadic, and unpredictable. Sclerotia from rust *(Claviceps purpurea)* have been found in assemblages of charred floral remains from Neolithic sites in

Central Europe (Knörzer 1971b), indicating that fungoid diseases would have posed a threat to Neolithic crops.

Lodging

Lodging is the laying down of a crop. As an ear matures, it becomes heavier and the stem becomes increasingly less able to support it. The ear's weight is further increased when wet, and the beating action of wind and rain can easily cause the crop to lodge (Scheffer 1972). Lodging can result in serious problems. First, lodged crops do not dry easily, and rots can set in. Second, the grain may not fill properly, and crops may ripen slowly (Gill and Vear 1980:57). Third, ripe grain may sprout before being harvested. Finally, harvesting is extremely difficult when the crop is lying flat. Lodging can occur at any time during the growing season, but the propensity to lodge increases as the crop ripens. Lodging may not pose the same devastating threats as do rusts, blights, and smuts; nonetheless, losses can be significant and some amount of lodging can be expected to occur every year.

Summary

Cereal crops are vulnerable to a variety of hazards throughout the growing season. Extremes of temperature and precipitation are perhaps the most important, for they can affect crop growth directly at any phase of development. Temperature and precipitation may provide conditions favoring the development of fungoid diseases, while severe storms contribute to lodging. A final hazard, not mentioned above, is that bird and mouse predation can substantially reduce crop yields. Having looked briefly at the general factors affecting crop yields, it is time to consider each major crop.

Wheat

An ear of wheat is composed of individual spikelets arranged above one another along a central axis known as the rachis. Each spikelet has a single set of glumes enclosing up to eight kernels, and each kernel is encased by its own lemma and palea. The number of kernels per node provides the primary characteristic for distinguishing between species of wheat. A second characteristics is the ease with which the kernels are released from their glumes. Wheats are described as being either "naked" or "hulled." Hulled wheats have strong glumes that tightly hold the kernels in their florets, whereas naked wheats are easily released. The obvious advantage of naked wheats is that when the crop is being processed, less labor is required to free the kernels from the ear. The advantage must be weighted against crop losses: should harvesting be delayed until the dead ripe stage, kernels of naked wheats may fall out of the ear. Moreover, because the grains of hulled varieties are hidden by the lemma, the palea, and the glume, hulled wheats are more resistant to fungoid diseases and to bird predation.

From a taxonomic viewpoint, wheats are divided into three classes according to their chromosome number. Diploid wheats have 14 chromosomes, tetraploid wheats have 28 chromosomes, and hexaploid wheats have 42 chromosomes. The chromosome number is relevant to the problem of Neolithic subsistence only in that the more chromosomes, the greater the genetic plasticity of the wheat. Thus tetraploid and hexaploid wheats are readily amenable to the breeding and development of varieties particularly suited to local soil and climatic conditions. Diploids, on the other hand, are less so.

Early Neolithic farmers planted two species of wheat, einkorn (*Triticum monococcum* L.) and emmer (*Triticum dicoccum* Shrank). During the very latest phases of the Early Neolithic spelt[1] (*Triticum spelta* L.), and a naked wheat (*Triticum aestivum/ durum*) appeared. Each is discussed below.

Einkorn

Triticum monococcum L. is a diploid hulled wheat with both winter and spring varieties; it is a homogenous species with a low morphological diversity. Einkorn is resistant to cold, heat, drought, fungoid diseases, and bird predation. It can grow well on poor soils, although it ripens later and has a lower yield than emmer, spelt, and naked wheat (Percival 1921:10; Peterson 1965:10). Einkorn occurs regularly at Neolithic sites, and at Rhine valley sites it is frequently more common than emmer (Knörzer 1980).

---

[1]*Triticum spelta* L. is often commonly referred to as "large spelt." Some confusion surrounds the use of the term "spelt," for *Triticum monococcum* has been referred to as "small spelt." Throughout the text, "spelt" refers to *Triticum spelta*.

Emmer

Triticum dicoccum Shrank is a tetraploid hulled wheat. It can be grown both as a winter and spring crop, but it appears to prefer spring planting in temperate climates (Percival 1921). Emmer thrives in a dry, prairie region with hot, short summers, and does not do as well in humid areas (Carleton 1901:6–10). Emmer is highly resistant to rusts and other fungoid diseases, but it does not withstand frosts (Percival 1921:188). Because it is a tetraploid wheat, emmer has moderate genetic plasticity and responds well to breeding. Emmer is present at all Early Neolithic sites, although it is not always the dominant wheat species.

Spelt

Triticum spelta L. is a hexaploid hulled wheat having both winter and spring varieties. Spelt is particularly hardy and can withstand continued frosts—and even snow—during the growing season. It does well in raw, cold climates and is highly resistant to fungoid diseases. Its yield is slightly lower than that of the naked wheats; but its hardiness offsets the low yields. The origin of the species is unclear; however, evidence suggests that it first appeared in the mountainous or upland region of Central Europe at some point during the late Early Neolithic. It is rare at Early Neolithic sites (see Willerding 1980).

Naked Wheat

Much controversy surrounds the classification of the polyploid naked that first appeared in Central Europe during the late Early Neolithic. Heer (1865) originally classified this wheat among the tetraploid naked wheats. Neuweiler (1905) questioned the validity of Heer's classification and placed it among the hexaploid naked wheats. More recently, paleoethnobotanists have reopened the controversy (Jacomet and Schlichtherle 1984; Kislev 1984). Although the precise taxonomic classification not yet been resolved, current evidence suggests that naked wheat is morphologically more similar to the tetraploid durum wheat, *Triticum durum* Desf., than it is to the modern-day bread wheat, *Triticum aestivum* L. Therefore, this prehistoric polyploid naked wheat is provisionally being referred to as *Triticum aestivum/durum* (see Jacomet and Schlichtherle 1983). Modern varieties of both durum and bread wheat have higher crop yields than emmer. Durum wheats tend to be hardier than bread wheats; consequently, their yields are higher in colder, damper weather than are bread wheat yields. Because the chromosome number of the hexaploid wheats is higher, bread wheats are more receptive to breeding; although both can be manipulated to withstand specific climatic and soil conditions. *Triticum aestivum/durum* is reported at Early Neolithic sites in the German Democratic Republic, northern areas of the Federal Republic of Germany, Poland, and Czechoslovakia. It predominates at Middle Neolithic sites (Willerding 1980).

Crop Processing

Harvesting takes place in the late summer, and it requires intensive labor over a relatively short period. Ideally, harvest should occur during hot sunny weather, but this is not always the case. When the crops are ripe, the harvest must take place immediately. Delays could result in losses due to the shattering of the ears during harvesting, or the kernels falling out in high winds (Zimmerman 1950). The heaviness of the ear makes the plant highly susceptible to lodging during storms, and if lodging occurs, the harvest must be delayed until the crop dries. Meanwhile, ripe kernels could germinate in the ear. Thus if the weather does not hold, sizable portions of the crop can be lost. The crop can be harvested by one of several methods (Hillman 1984). Ears can be broken off just below the base, several culms can be held together and cut at some point along the straw, or the entire plant can be uprooted. Each of these methods has drawbacks.

Ear-by-ear plucking is time-consuming, but it ensures that few weeds are collected (Knörzer 1971b) and results in relatively weed-free seed grain for the next crop. Furthermore, the whole ear can be stored for processing immediately prior to use, or the crop can be cleaned and the clean kernels stored. Moreover the standing straw can then be harvested for other purposes, such as: thatch, bedding, or cattle fodder. The second and third methods require less work in the field; consequently the harvest can be brought in faster. With these methods, weeds growing next to or climbing up the plants will be harvested along with the crop. When this happens, crop cleaning requires more time and a greater number of weed seeds are incorporated in the seed grain for the next crop. If the entire plant is uprooted, quantities of dirt, grit and sand may be introduced into the stored grain.

Crop cleaning entails three basic steps. The ears must be detached from the straw; the spikelets must be separated from the ear; and the kernels must be first freed and then separated from the rachis and chaff. Four distinct activities are used to clean the grain. Threshing separates the ears from the straw. In the case of naked wheats and barleys threshing frees the kernels and chaff from the rachis. But with emmer and einkorn, additional pounding or grinding is needed to release the grain from the spikelets (Hillman 1984). Sieving is then needed to separate plant parts by size so that kernels are free from larger pieces of chaff, straw, and plant debris. Finally, the partially cleaned crop must be winnowed to separate the lighter chaff and plant debris from heavier debris and kernels.

Emmer, einkorn, and spelt require threshing, pounding, winnowing, and sieving. Naked wheats require only threshing, winnowing, and sieving. Storage can occur at virtually any step of the process. Sheaves of harvested grain can be stored, although they are bulky, and require a large amount of space. Neolithic houses were not tightly built, and there were no domestic cats. While in storage, sheaves of grain would have been susceptible to mouse predation. Losses could have been dramatically reduced by storing the grain in ceramic vessels. Entire ears could have been stored in vessels, but ears are bulky. Grain was probably stored after the first or second sieving, when most of the extraneous debris had been removed.

Prehistoric Crop Yields

As mentioned earlier, climatic conditions are probably the most important factor affecting crop yields. Reconstructions of annual weather patterns provide the best approach to evaluate year-to-year harvest fluctuations. Changes in the annual growth rings of trees are one vehicle through which yearly climatic variation may be identified. The key to using dendrological data to derive climatic information lies in using a species that is sensitive to climatic fluctuation. With such a species, weather extremes are responsible for variations in the width of annual growth rings. Extreme climatic conditions result in slower growth and trees form narrow annual growth rings, whereas under optimal growing conditions trees produce wide growth rings. In principle, fluctuations in the yearly width of annual growth rings should correlate with the severity of climatic fluctuations. Other factors, such as insect predation and the amount of available sunlight, also influence the development of growth rings, and in practice it is not always possible to correlate width variation with climatic fluctuations.

Oaks (*Quercus robur* L. and *Quercus petraea* (Mattuschka) Liebl.) are the most commonly used species for dendrochronological studies from archaeological sites. During the Atlantic period, Central Europe provided the optimal ecological conditions for the growth of oaks. Variation in the width of annual growth rings can be attributed to a variety of causes, including site conditions, insect and animal predation, and tree disease—as well as weather conditions during the growing season. Attempts to correlate nineteenth century crop yields with nineteenth century oak tree rings have proved unsuccessful.[2] Fluctuations in historic crop yields, therefore, provide the best alternative for modeling variability in prehistoric crop yields. Selecting reference years for modeling prehistoric crop yields depends on general climatic conditions, the availability of quantified data on crop yields, and the use of unimproved wheats.

As discussed in the first chapter, Early Neolithic villages appeared in Central Europe during the Atlantic climatic period. This was the warmest period to have followed the retreat of the Würm glacier. Temperatures were 1°C to 2°C warmer than at present, with summer temperatures averaging 17.8°C and winter temperatures averaging 5.2°C, as compared to the present summer average of 15.8°C and winter average of 4.2°C (Dansgaard 1984; Lamb 1977). Dansgaard (1984:205) argues these averages may be somewhat misleading. Noting that fluctuations between warm and cold phases were constant in amplitude, he suggests the Atlantic differed only in duration—and not intensity—from warm phases that have occurred historically. Flohn and Fantechi (1984:1) concur, noting that seasonal mean temperatures can vary from 8 to 13 degrees in winter and from 5 to 7 degrees in

---

[2]I am indebted to Dr. B. Becker of Universität Hohenheim for kindly allowing me to use his published and unpublished nineteenth century annual growth ring data to attempt the correlation and for discussing specific problems of Central European dendrochronology with me.

summer; moreover, long-term averages spanning 30 years often vary from 1°C to 2°C. If these arguments are accepted, then it can be argued that historic and prehistoric crop yields may be expected to show similar intensity of year-to-year variation, even if the absolute yields are slightly higher or lower.

From A.D. 1550 until 1700 the Little Ice Age occurred in the Northern Hemisphere, and climatic recovery was not complete until the mid-1800's. The Northern Hemisphere has experienced a general warming trend since then, with the warmest phase beginning in 1916 and continuing to the present (Lamb 1977). During much of that time, normal farming activities were disrupted because of wars. Since 1950 the application of chemical fertilizers, herbicides, and pesticides in conjunction with the growing of improved bread wheats, and highly mechanized farming, have resulted in high crop yields. Thus the years from 1916 until the present are not suitable for comparative purposes.

The period from 1850 to 1900 was relatively warm; chemical fertilizers (other than manure), pesticides, and herbicides were not used; improved varieties of modern bread wheats were not common; and both planting and harvesting were done by manual labor. The 1850–1900 period was used as the reference years for determining crop yields. Statistical yearbooks from Württemberg, Federal Republic of Germany (Statistisch-Topographisches Bureau 1850–1905) provided annual data on the kernel and straw yields of pure stand einkorn, and of mixed cropped fields of emmer and spelt, as well as the yields of lentils, peas, and flax. In addition, commentaries provided information on specific factors affecting crop yields, and annual meteorological reports gave detailed information on weather conditions throughout the year.

There are several difficulties in using these data. First, reporting methods varied throughout the period; however, the adoption of the metric system in the 1870's resolved the problem of using different scales of measurement. Second, some yields were often reported as a total harvest combining both spring and winter cereals, rather than being reported separately. Other years reported only yields from winter crops. It was possible to extract the yields of spring and of winter wheat crops for 1880–1891, and these years have been used to estimate a mean yield for the crops. One further problem should be mentioned. Farmers sowed emmer and spelt as a mixed crop because they felt that when sown together, a more predictable yield resulted. Since spelt is hardier than emmer, the mixed emmer-spelt crop probably showed less annual variation than emmer would have shown, had it been grown as a pure stand crop. Thus yields of the mixed emmer-spelt yields may not show as much yearly fluctuation, and standard deviations may not be as wide as could be expected prehistorically.

According to the statistical yearbooks an estimated 200 kg/ha are lost to mice and fungus or are needed as seed for the following year's crop. When the amount is subtracted from the mixed emmer-spelt winter mean of 1045 kg/ha, an average yield of 845 kg/ha is obtained. This figure is only slightly larger than the estimate of 800 kg/ha Bakels suggests be used as an estimate for yields of prehistoric emmer (Bakels 1982:10), while the adjusted yield of 556.80 kg/ha for spring emmer-spelt is substantially below the figure used by Bakels. Table 3 shows the mean yield per hectare and the standard deviation for the straw and kernels of both wheat crops.

Significance of Cereals to the Human Diet

Cereals are important for two reasons. First, they are easily stored and, when properly stored, little loss of nutrients occurs for a period of several years (Aykroyd and Doughty 1970:34). Second, cereals are excellent sources of energy. Wheat provides 3300 calories per kilogram (Watt and Merrill 1975) primarily as carbohydrates. Cereals contain biologically incomplete proteins. In particular, they have low levels of lysine and isoleucine. Vegetable proteins with complementary amino acid patterns must be eaten in conjunction with cereals, or complete proteins must be obtained from animal sources in order to maintain adequate nutritional levels. The next section considers vegetable proteins.

Legumes

Two legumes, peas and lentils, were cultivated during the Neolithic but their role in the Neolithic subsistence system has been overlooked in favor of cereal cultivation. Legumes provide one of the best sources of plant proteins (Table 4), but they both lack tryptophan and methionine, essential amino acids. The deficit complements wheat, which has high levels of both; furthermore, wheat has low levels of isoleucine and lysine, two amino

Table 3

Historic Yields of Einkorn and Emmer-Spelt
(kg/ha)

Crop	Kernels		Straw	
	Mean	Standard Deviation	Mean	Standard Deviation
Winter Einkorn	835.20	134.70	2153.00	288.79
Spring Einkorn	645.00	74.80	2023.50	212.79
Winter Emmer-Spelt	1045.00	189.90	2591.50	359.73
Spring Emmer-Spelt	756.80	98.90	2187.30	169.96

Source: Statistisch-Topographisches Bureau 1850–1905

acids that occur in high levels in both peas and lentils. Thus peas and wheat, or lentils and wheat, are complementary proteins. A Neolithic diet of cereals and legumes would have provided sufficient protein for growth.

Legumes require large amounts of nitrogen for vigorous growth. Unlike most plants, legumes are not dependent on nitrogen compounds in the soil, because legumes and a nitrogen-fixing bacteria, *Rhizobium radicicola*, have formed a symbiotic mutualism (Gill and Vear 1980:139). *Rhizobium radicicola* invades the root hairs of legumes and there stimulates the host to produce nodules in which the bacteria lives. *Rhizobium radicicola* derives its carbohydrates from the host plant. In exchange it provides the legume with nitrogen; thus legumes are not limited by the availability of soil nitrogen. When the roots decay after the death of the plant, the bacteria and nitrogen are released into the soil. Nitrogen fixation in the soil can be as low as 80 kg/ha for peas and as high as 600 kg/ha for clovers (Gill and Vear 1980:140).

Cereals and legumes can be interplanted, but the purpose of such inter-cropping is more to provide the legumes climbing support than for the wheat to derive any benefits from the relationship between *Rhizobium radicicola* and the legume. As mentioned above, nitrogen created by *Rhizobium radicicola* enters the soil only after decay of the root nodules has occurred.

Table 4

Nutritional Value of Peas and Lentils
(per 100 grams uncooked legumes)

	Calories	Protein (g)	Fat (g)	Carbo-hydrate (g)
Lentils	340	24.7	0.9	19.3
Peas	348	24.2	1.0	62.7

Source: Watt and Merrill 1975

Peas

Peas (*Pisum sativum* L.) are a cool-weather crop with a relatively short growing season. Fields should be used for peas only once every six or seven years (Zimmerman 1950:161) and approximately 100 kg/ha seed are sown. Shelled peas can be stored for five to six months in the winter (Wasserman 1967). As with the wheats, historic records from Baden-Württemberg were consulted to obtain figures for average yields (Statistisch-Topographisches Bureau 1850–1905). Table 5 is based on harvests from 1880–1891, the same years that the most complete wheat records were available.

Lentils

Lentils (*Lens culinaris* Medik.) have a life history similar to peas. They prefer cool weather and have a relatively short growing season. In mild areas, lentils can be planted with winter barley. Lentils are planted in early April at 80 kg/ha (Zimmerman 1950:164), and decreased yields will result from delays in planting. Manuring also results in decreased yields, for it promotes a leafy growth rather than seed production. Lodging can be a problem in lentils, and they are very poor competitors: lax weed control can reduce yields significantly (Hawtin, Singh, and Saxena 1980:618). Lentils ripen before peas, and the crop must be harvested before the pods dry; otherwise the pods can shatter and losses may be high.

Table 5

Historic Pea and Lentil Seed Yields
(kg/ha)

Crop	Mean	Standard deviation
Peas	1473.40	579.06
Lentils	1152.00	511.56

Source: Statistisch-Topographisches Bureau 1850–1905

Oil Plants

Two oil plants are occasionally found at Early Neolithic sites. These are opium poppy (*Papaver somniferum* L.) and flax (*Linum usitatissimum* L.). Seeds from both plants are eaten often in Central Europe, but rarely in English-speaking countries; consequently, the importance of these crops is often overlooked by Anglo and American prehistorians. Oil producing plants are important because they provide a domestic source of linoleic acid, a fatty acid the body cannot synthesize. Both poppy seed and linseed have high levels of linoleic acid.

Poppy

Papaver somniferum L. is best known because latex collected from slits in the unripe capsule is used to make opium. Poppy seeds have been found at Early and Middle Neolithic sites, although to date no finds of scored, unripe poppy capsules have been found. The possibility that opium may have been manufactured cannot be eliminated, but currently there is no evidence for its production. Poppy would have been an important source of oil, for its seeds contain 40–55% oil (Vaughn 1980:195) with approximately 65% linoleic acid and 20% oleic acid (Gill and Vear 1980:194). The seeds could have been eaten whole, or they may have been pressed, and 1,000 kg of seeds deliver approximately 350 kg oil.

Poppy grows best after legumes in a crop rotation. Poppy is seeded at 3–4 kg/ha. It germinates very slowly, so a weed-free soil is needed (Zimmerman 1950:207). Poppy ripens after the cereal harvest in August. Several plants can be bundled together and harvested like cereals. The heads are then allowed to dry before being threshed (Zimmerman 1950:207).

Linseed and Flax

Flax (*Linum usitatissimum L.*) is cultivated both for linseed, its seed, and for flax, its fiber. Linseed is rarely eaten in the U.S., Canada, or England, but it is commonly consumed in Central and Eastern Europe. In fact, in the folk taxonomy of health foods, linseed fulfills the same role in Central Europe that sunflower seeds play in parts of the United States. Flax contains 35–40% oil (Hoffman 1961:264) and 20% protein (Gill and Vear 1980:196; Vaughn 1980:141). Linseed meal (the remains from oil pressing) could have been fed to the pigs. In addition, flax stems contain fibers that, once extracted, can be spun and then woven into linen. *Linum usitatissimum* is now cultivated for two distinct purposes: one variety is for flax production, the other is for seeds. High seed yields currently are not compatible with good flax production (Gill and Vear 1980:198), it is unclear whether the dichotomy existed prehistorically.

Flax thrives in a warm, damp climate with even rain throughout the growing season, and it does well on most soils other than on peats, clays, or sands (Zimmerman 1950:208). Flax is seeded at 140 kg/ha with planting beginning in late March or early April and lasting until June (although crops planted in April are usually 20% larger than those sown in June) (Hoffmann 1961:300). Harvest takes place in late August or early September. Plants are pulled out by their roots and raked to disengage the capsules from their stems. The capsules are then threshed to obtain the seed, and the stems must be decomposed to obtain the fibers.

Domestic sources of oil were important for contributions of linoleic acid to the diet. The high oil content of both poppy seed and flax would have made them welcome additions to the Neolithic diet. That each of these plants may have provided an additional resource would have only amplified their value.

Cereals and Legumes in Neolithic Diets

The foregoing discussion briefly presented key crops grown during the Early Neolithic. Wheat would have been cultivated as a source of carbohydrates, while peas and lentils would have provided a much-needed complement to cereal proteins. Both poppy seed and linseed would have provided a source of vegetable oils to the diet.

There is no inductive way of determining the relative importance of cereals, legumes, and oil plants in the Neolithic diet. Ethnographic studies of populations of cereal cultivators in India (Sen Gupta 1980) suggest the importance of cereals in the diet of each group varied, with cereals providing from 62% to 92% of the daily calories. This amounts to each person's consuming between

1200 and 1800 grams of cereals daily. Lentils were not as significant; nonetheless, among groups who cultivated pulses, each person consumed from 17 to 23 grams daily. The figure provides an arbitrary guideline for modeling prehistoric strategies. Before considering the ways that crops may have been combined into a Neolithic subsistence strategy, planting and cropping strategies must be identified so that annual crop yields can be established. The life histories and ecological requirements of crop weeds provide one of the best vehicles for evaluating planting and cropping strategies as well as field conditions.

Weeds

A weed is any plant growing where it is not wanted; however, weeds are particularly characterized by their unwanted appearance in fields or gardens. Thus the definition of weeds is based primarily on aesthetic and economic considerations rather than biological characteristics. Weeds compete with crops for available water, light, and nutrients, and the competition can result in decreased crop yields. Weed seeds may also contaminate seed grain and foodstuffs. The following discussion examines the life history of weedy plants, identifies the utility or disutility of "weeds," and considers the types of information that may be derived from the presence of particular species in archaeological assemblages.

Life Cycle

Weeds are divided into three classes. *Annuals* complete their life cycle and produce seeds within one year. This broad class is further subdivided into winter and summer annuals. Winter annuals germinate in the fall and produce seeds in early summer, while summer annuals germinate in the spring and produce seeds in the fall. *Biennials* require two years from germination to flowering and seed production, whereas *perennials* live for more than two years. Weeds are thus classified according to whether they live one, two, or more than two years. But their behavior is not constant from region to region: annuals in severe climates may be biennials or perennials in mild climates (Muenscher 1980).

Weeds propagate by seeds and vegetatively. Annuals and biennials produce tens of hundreds or tens of thousands of seeds per plant. Unlike annuals and biennials, perennials reproduce vegetatively as well as through seeds. The stems of perennials can creep along the ground surface (stolons), or they can be buried in the earth (rhizomes, tubers, bulbs, and corms) to a depth of several meters. Such perennials can be particularly difficult to eradicate, for although the top of a plant may be destroyed, regrowth can occur from underground parts. Weeds rely on a number of transporting or scattering agents, including wind, water, animal movement, and humans. Some are extremely light and are easily transported by the wind. Others have sticky substances or small hairs that cause them to adhere to skin, clothing, fur, or feathers, thereby allowing them to be carried to new areas.

Still others are eaten by herbivores when grazing, and may pass undamaged through the digestive tract of an animal (Table 6). More than 75 species of European weeds have been found in cattle dung (Heintz, cited in Ridley 1930:339), with up to several thousand seeds of individual species being found in 1000 kg of livestock dung (see Table 7). Harm and Keim (1934, cited by Robbins 1952:45) argue that of 1000 weed seeds fed to livestock, "an average of 6.7% viable seeds was recovered." The survival of viable seeds in livestock digestive tracts allows livestock to consume seeds in one area and transport them elsewhere. Grazing livestock on crop stubble could thus facilitate the introduction of new (or re-introduction of old) weedy species to plots of cropland.

Weed seeds can also be dispersed through the use of barnyard manure as crop fertilizer. Several studies (Harm and Keim 1934; Pammel and King 1909, as cited in Robbins 1952:45) suggest that, although the viability of weed seeds decreases proportionally to the amount of time manure is composted, seeds in manure can remain viable for up to six months. Feeding crop-cleaning debris to livestock can present a problem. Threshing and winnowing debris can contain thousands of weed seeds per kilogram of debris, many of which can pass through the digestive tract of cattle, sheep, goats, and pigs, and may be incorporated in manure (Table 8). Unless composted for six months, manure re-introduces weeds in crop fields every time it is applied.

Table 6

Seed Germination After Passing Through Livestock Digestive Traact

Species	Cow	Pig	Sheep
Agropyron repens Quack grass	+	−	+
Bromus secalinus Rye grass	−	−	−
Chenopodium album Lamb's quarters	+	+	+
Daucus carota Wild carrot	−	−	−
Galium aparine Cleavers	−	−	−
Polygonum convolvulus Wild buckwheat	−	−	−
Rumex crispis Curly dock	+	+	+
Setaria viridis Green foxtail	+	+	+

Source: Muenscher 1980 + germinated −did not germinate

Crop Competition

Studies of weed populations in arable soil show that millions of weed seeds are present per acre. Brenchly and Warrington (1930) demonstrated that 45 million weed seeds were present per acre. (Their figure was derived only *after* excluding a further 113 million seeds from corn poppy (*Papaver rhoeas*), because of a par-

Table 7

Viable Seeds per 1000 kg Dung

Plant Species	No. of Viable Seeds		
	Cow	Sheep	Pig
Chenopodium album Lamb's quarters	4880	9240	4441
Polygonum persicaria Lady's thumb	6164	n.d.	n.d.
Polygonum lapathifolium Pale persicaria	687	1327	1684
Galeopsis tetrahit Hemp nettle	2524	n.d.	n.d.

Source: Kormso 1930

ticularly heavy poppy infestation at the time the inventory was taken.) In Finland one study showed that 550 weed plants were present per square meter, along with an almost equal number of cereal tillers (Erviö 1972), although another study (Laursen 1971) found that the number of weed plants varied throughout the growing season.

The competitive ability of crops is determined by a number of factors. First competition is most serious during the first four to six weeks after germination (Robbins 1952). Second, the density of crops affects the number of weeds. Cereals appear to outcompete weeds in the earliest stages of growth. The more densely a field is sown with cereals, the less room there is for weeds. Erviö (1972) shows that at a wheat-seeding rate of 26 kg/ha, weed growth smothered cereal growth. But at seeding rates of 200 and 400 kg/ha, the cereals effectively reduced weed competition. The ability of dense cereal stands to reduce weed yields has also been noted by Brenchly and Warrington (1917). Third, weeds are more likely to compete successfully if their growth habits are similar to

Table 8

Weed Seeds in Winnowing and Threshing Debris

Species	Threshing		Winnowing	
	No. Samples (1 kg)	Average No. of Seeds	No. Samples (1 kg)	Average No. of Seeds
Chenopodium album Lamb's quarters	n.d.	116,798	n.d.	15,814
Galium aparine Cleavers	10	8,652	n.d.	n.d.
Lapsana communis Nipplewort	4	16,875	2	200
Polygonum convolvulus Wild buckwheat	16	9,242	n.d.	1,100
Polygonum lapathifolium Pale persicaria	30	43,273	6	15,233
Polygonum persicaria Lady's thumb	6	20	653	n.d.
Rumex acetosella Sheep's sorrel	11	619	n.d.	20
Rumex crispis Curly dock	n.d.	3,000	n.d.	n.d.
Sinapsis arvensis Mustard	8	40,131	3	2,167

Source: Kormso 1930

the crop they infest. Moreover, weeds closely imitating the crop stand a better chance of not being eradicated during weeding or hoeing.

In addition to competing with crops, weeds can contribute to their contamination. On the one hand, weeds can act as hosts to crop diseases, such as those listed in Table 9. On the other, seeds from weeds can also contaminate flour. Their effects can vary from causing bread or gruel to have an unusual taste or color to adding poisons to the flour.

Table 9

Weed Hosts to Crop Diseases

Weed Host	Disease	Cause	Crop
Agropyron spp.† Quack grass	Black stem-rust	*Puccinia graminis*	Wheat Barley
Avena fatua† Wild oat	Black stem-rust	*Puccina graminis*	Wheat Barley
Agropyron spp.† Quack grass	Ergot	*Claviceps purpurea*	Barley
Berberis vulgaris† Barberry	Black stem-rust	*Puccinia graminis*	Wheat Barley
Self-sown cereals‡	Yellow rust	*Puccinia striiformis*	Wheat Barley
Self-sown cereals‡	Brown rust	*Puccinia hordei*	Barley
Self-sown cereals‡	Powdery mildew	*Erysiphe gramminis*	Wheat Barley
Poa trivialis‡ Rough meadow grass	Leaf spot	*Septoria tritici*	Wheat

Sources: †Muenscher 1980: Table VII; ‡Roberts 1982:273–274

Weed Control

Weeds are controlled by interrupting their life cycle. With annuals and biennials, the best control is to prevent them from seeding. Hand-pulling and hoeing, when weeds are in their seedling stage, are the most effective methods. Mowing or grazing livestock on the plot can also be effective, if done before the weeds flower and seed. As mentioned above, if grazing occurs after the weeds have produced seeds but not scattered them, animals may transport viable seeds to other fields.

Crop rotation can also be effective in reducing weed infestation. Different weed communities adapt themselves to the various crops. When crops are rotated, weed communities do not have time to develop. If spring-sown crops follow winter crops with a winter fallow in between, then weeds that germinated in the fall are killed in the spring when the plot is cultivated. Similarly, if winter crops follow spring-sown crops, the winter crop smothers the spring annuals that otherwise would have appeared.

It is much more difficult to control perennials. Rhizomes, corms, and tubers store nutrients that allow perennials to reestablish themselves, even if the top of the plant is destroyed. Because they can be deeply buried, rhizomes, corms, and tubers are extremely difficult to eradicate. Other than using herbicides, the best method is to repeatedly destroy the above-ground structures until the underground food reserves are exhausted. Eradication requires repeated hoeing and hand-pulling at close intervals throughout the growing season.

Recurring Weeds at Neolithic Sites

Inventories of weeds found in association with cereal and flax crops suggest that although as many as fifty species of weeds occur, approximately half occur only sporadically (Brenchley and Warrington 1930, Erviö 1972). Of the remaining species, less than half recur consistently in high frequencies. Such is also the case for the occurrence of prehistoric weed seeds in the archaeological assemblage. One problem in attempting to synthesize findings from Early Neolithic sites is that data have often been collected by incomparable methods. For example, reports for some sites are limited only to cereal impressions in pottery. Others consider only charred materials that were visible to the naked eye during the course of excavation. Still others include only charred material from soil samples that were water-sieved through a nested set of screens. Willerding (1980) synthesized archaeobotanical reports from 100 Early Neolithic sites throughout Central Europe. Although floral materials were collected using a variety of techniques, Willerding was able to conclude that thirteen species were characteristic of Early Neolithic sites, and that a further sixteen occurred less frequently.

In order to derive a comparable estimate of the frequency of occurrence of weedy species at Early Neolithic sites, archaeological botanical reports from fourteen sites were examined (Bakels 1984; Bakels and Rousselle 1985; Gregg 1984; Knörzer 1971a, 1980). Recovery methods for each of these analyses entailed water screening soil samples through a nested set of screens, the smallest mesh size being 0.5 mm. Charred botanical remains were identified using a dissecting microscope.

Weedy species present at a minimum of eight sites have been labeled as commonly occurring species. Species found at four or more sites are considered to be frequently occurring species, while those present at fewer than four sites are considered to be rare (Table 10). Of the latter group, fourteen are reported at only one site. A brief review of the life history and potential economic value is discussed for each of the most common species, and is summarized in Tables 11–12. Similar information for the frequently and for the rarely occurring species are summarized in Tables 13 and 14 respectively.

Commonly Occurring Weeds

Wild buckwheat (*Polygonum convolvulus* L.) is among the most common non-domesticated weed seeds in the floral assemblage of Early Neolithic sites. Wild buckwheat is a creeping spring annual that prefers a damp soil along sunny stream banks and is also one of the most persistent weeds of cultivated fields. It commonly occurs in wheat, barley, flax, and peas (Fabricius and Nalewaja 1968; Friesen and Shebeski 1960; Gruenhagen and Nalewaja 1969, Hume *et al.* 1983). Wild buckwheat twines on the crop it grows with, and competes directly with it for available nitrogen, phosphorous, and water (Fabricius and Nalewaja 1968). Losses of 1% to 26% have been reported as a result of wild buckwheat infestation in cereal fields (Hume *et al.* 1983), and losses of over 20% have been reported for flax fields (Friesen and Shebeski 1960: Table 4).

Table 10

Weedy Species Found at Early Neolithic Sites

Commonly Occurring	Frequently Occurring	Rarely Occurring	
Bromus secalinus	*Echinochola crus-galli*	*Atriplex* sp.	*Chenopodium glaucum*
Rye grass	Cockspur	Orache	Glaucous chenopodium
Bromus sterilis	*Galium spurium*	*Chenopodium hybridum*	*Papaver dubium*
Sterile brome	False cleavers	Sowbane	Long headed poppy
Bromus sp.	*Polygonum persicaria*	*Polygonum lapathifolium*	*Polygonum calcatum*
Brome grass	Lady's thumb	Pale persicaria	Low knotgrass
Chenopodium album	*Rumex* sp.	*Poa* sp.	*Sinapsis* sp.
Lamb's quarters	Dock	Meadow grass	Mustard
Lapsana communis	*Vicia hirsuita*	*Rumex acetosella*	*Rumex sanguineus*
Nipplewort	Hairy tare	Sheep's sorrel	Red veined dock
Polygonum convolvulus	*Phleum nodosum*	*Rumex acetosa*	*Solanum nigrum*
Wild buckwheat	Cat's tail	Sorrel	Black nightshade
		Setaria sp.	*Setaria viridis*
		Bristle grass	Green foxtail
		Silene nutans	*Silcne vulgaris*
		Nottingham catchfly	Bladder campion
		Stellaria graminea	*Melandrium rubrum*
		Lesser stitchwort	Red campion
		Trifolium sp.	*Trifolium repens*
		Clover	White clover
		Veronica hederifolia	*Veronica arvensis*
		Corn speedwell	Wall speedwell

Table 11

Characteristics of Commonly Occurring Weeds

Species	Growth Habit	Propagation	Seeds per plant	Seedlings Appear	Survival in Livestock Digestive Tract	Long-term Viability in Soil	Places in Which Weed Appears	Means of Control
Polygonum convolvulus Wild buckwheat	spring annual	seed	15,000–30,000	Flush May-June; all summer	excellent	decades	wheat, barley, flax	sow clean seed; hoe
Chenopodium album Lamb's quarters	spring annual	seed	72,450	spring flush; small fall flush	good	decades	wheat, barely, poppy	hoeing and pulling
Bromus secalinus Rye grass	winter annual	seed	1,450	fall	good	poor	winter wheat	sow clean seed
Bromus sterilis Sterile brome	winter annual	seed	n.d.	fall	good	very poor	waste places; shadowed meadows	sow clean seed
Lapsana communis Nipplewort	spring annual	seed	650	spring flush	n.d.	poor	hedges, crops, gardens	prevent from seeding

Sources: Georgia 1938; Gross 1924; Holm *et al.* 1977

Wild buckwheat is a prodigious seed producer, and its seed coat is thick. Buried seeds remain viable for decades and can withstand storage in manure (Holm *et al.* 1977). The seed is large and heavy, so unless eaten by livestock along with plant foliage, seeds drop directly to the ground, where they lie dormant throughout the winter. The majority of seedlings appear in the late spring, although they continue to emerge all summer. Seeds ripen throughout the mid- to late summer, and there is no uniform "harvest" time. Hoeing and weeding can reduce infestation, but because wild buckwheat germinates and produces seeds during the summer, these activities must continue throughout the growing season. The large population of buried seeds and the length of their viability make it extremely difficult to eradicate. In fact, wheat-wheat-fallow and wheat-fallow rotations over a twenty year period had little effect on wild buckwheat densities (Hume 1982).

The vegetative portions of wild buckwheat are low in palatability for livestock, and, in comparison to other weeds, its leaves and stems are low in protein, oil, and fiber (Holm *et al.* 1977:183). Such is not the case for the seeds of wild buckwheat; its seeds are nutritious for human populations (Table 12). Perhaps most significant, the achenes have a high lysine content, and their amino acid patterning complements that of cereals. Bertsch (1954) reported stores of wild buckwheat in ceramic containers at Ehrenstein, an early Middle Neolithic village in Southwest Germany, and charred achenes are common at most Early Neolithic sites (see Knörzer 1971a: Table 1). My experiments show the collection of wild buckwheat achenes was time-consuming, and that low yields were obtained because of the uneven ripening time of the seeds. Wild buckwheat may have been used from time to time as a starvation food, or as a condiment to enliven the taste of gruel, although there is no evidence for its intentional cultivation.

Lamb's quarters (*Chenopodium album* L.) is a pioneering annual commonly found on bare ground, in fields and gardens, and along shorelines and pathways. It produces an average of 72,450 seeds per plant (Stevens 1932). The seeds are heavy, so unless eaten by livestock and transported elsewhere, they drop near the parent plant. While germination occurs primarily in the spring, seedlings continue to appear throughout the summer, and a second peak occurs in mid-summer (Ogg and Dawson 1984). Lamb's quarters can be poisonous to sheep and pigs if eaten in large quantities over a long period of time (Herweijer and den

Table 12

Essential Amino Acid Pattern of Weed Seeds
(A/E ratio per gram protein)

Amino Acid	Hen's egg	Wild buck-wheat	Field dock†	Green foxtail	Lamb's quarters	Wheat
Isoleucine	129	123	111	96	105	106
Leucine	172	182	181	321	177	201
Lysine	125	125	126	34	142	72
Phenylaline	114	114	122	122	108	149
Tyrosine	81	61	68	62	76	85
Total sulphur-containing amino acids	107	120	100	124	117	124
Cystine	46	53	52	44	61	73
Methioine	61	67	48	80	56	51
Threonine	99	99	104	87	99	84
Tryptophan	31	27	34	33	47	45
Valine	141	150	154	121	130	135
E/T ratio‡	3.22	2.44	2.29	2.83	2.00	2.06
Amino acid score§	100	75	79	27	81	58

Source: Tkachuk and Mellish 1977
†*Rumex pseudonatronatus* L.
‡The E/T ratio is the proportion of total protein nitrogen formed by the essential amino acids.
§Expressed as an index of the hen's egg score.

Houter 1970). Lamb's quarters seeds are nutritious, and they can be ground into a flour having a nutritional value comparable to that of wheat flour. The seeds are known to have been used as a starvation food in medieval and historic times in Europe (Maurizo 1927). Whether lamb's quarters seeds were used by prehistoric populations in Central Europe remains to be demonstrated (Helbaek 1960), although Knörzer (1971c) suggests that the leaves were used as a salad.

Several brome grasses (*Bromus* sp.) occur in fields, pastures, gardens, and along pathways. They are spring annuals that reproduce by seed. A number of closely related species exist (Gill and Vear 1980) and two occur regularly at Early Neolithic sites: rye brome (*Bromus secalinus* L.) and sterile brome (*Bromus sterilis* L.). In general, brome grass seeds do not germinate after more than five years in the soil, and they are viable after passing through livestock digestive tracts.

Rye grass (*Bromus secalinus* L.) is a winter annual that often infests winter cereals. The plant mimics wheat, and it can be difficult to distinguish at the seedling stage. The glumes hold the grain fast in its ear, and if the plant is harvested along with the cereal crop, the seeds will be threshed and winnowed along with the crop. Prehistoric grains are as large as emmer and einkorn wheat, and it would have been an arduous task to separate rye grass from the domestic cereals. Knörzer (1967) suggests that rather than being an uninvited weed, rye grass was specifically harvested as a food. Rye grass occurs in high frequencies throughout the Early Neolithic, and is uncommon in floral assemblages thereafter. Sterile brome (*Bromus sterilis* L.) is a winter annual that is abundant in field hedges rather than in cultivated plots (Gill and Vear 1980:167). When young, the plant is palatable for livestock.

Nipplewort (*Lapsana communis* L.), a pioneering annual that is a member of the daisy family, is a shade-tolerant plant found in gardens and fields as well as in hedges and thickets. The "seeds" are actually small achenes with a tuft of hairs that allow them to be transported widely by the wind. After wild buckwheat, lamb's quarters, and rye grass, nipplewort is the most common wild plant represented in Early Neolithic floral assemblages. Groenman-van Waateringe (1971), noting the presence of this shade-tolerant species, suggests that trees shadowed the fields and argues that the fields were small.

Frequently Occurring Weedy Species

Cockspur (*Echinochloa crus-galli* (L.) Beauv.) is a low spring annual found in grain fields, farm yards, and on the edge of ditches and ponds (Oberdorfer 1979). A flush of seedlings appear in May, but germination can occur throughout the summer. The number of seeds cockspur produces depends on the number of tillers the plant produces. On a world wide basis, seed production varies up to 40,000 seeds per plant (Holm *et al.* 1977). Klingmann (1961) suggests an average of 7160 seeds per plant, and Kormso (1930) suggests that in Central Europe the plant produces between 200 and 1000 seeds. Higher seed production appears to be related to high levels of soil moisture, and cockspur infestations seem to be greater in wet fields.

Cleavers (*Galium* sp.) are represented by two species. False cleavers (*Galium spurium* L.) are reported frequently at Early Neolithic sites. The species is considered by some to be a subspecies of bedstraw (*Galium aparine* L.). Both have similar growth habits and ecological requirements. The plants are distinctive. They have square stems, their leaves occur in whorls, and small, bristly, hooked hairs cover the plant and seeds. The seeds cling tightly to clothing, fur, hair, shoes, and tools. False cleavers prefer damp soils. It is common in cereal crops and appears occasionally in flax fields (Oberdorfer 1979:739). Bedstraw occurs in meadows, cereal crops, hedges, and woodland borders. Cleavers have been used in cheese making; the sap has been used as a herbal remedy to treat skin diseases; and an infusion of the leaves in water has been used to alleviate digestive problems (Willfort 1982:309).

Lady's thumb (*Polygonum persicaria* L.) is a pioneering annual that grows to a height of 40–100 cm. The plant is found in crop fields, in gardens, in gullies, and along shorelines. The seeds are hardy. They have a good long-term viability in the soil, and large quantities of viable seeds have been found in animal dung. The plant can be used to produce a yellow-brown dye, and it has been used in herbal remedies (Willerding 1980:195). Whether it was used for these purposes during the Early Neolithic remains to be demonstrated.

Dock and sorrel (*Rumex* sp.) are pioneering perennials found in fields and pastures. Most species have a long taproot that enables the plant to regenerate after being mown or grazed, but dock and sorrel normally reproduce by seed. Sorrel and dock plants can produce up to 60,000 seeds per year, and buried seeds

Table 13

Characteristics of Frequently Occurring Weeds

Species	Growth Habit	Propagation	Seeds per Plant	Seedlings Appear	Survival in Livestock Digestive Tract	Long-term Viability in Soil	Places in which Weed Appears	Means of Control
Echinochloa crus-galli Cockspur	spring annual	seed	7,160	all summer	n.d.	n.d.	waste places; spring cereals	sow clean seed
Galium aparine Cleavers	winter annual	seed	350	fall flush; all winter	good	very poor	peas, lentils, cereals, flax	rake from crop
Polygonum persicaria Lady's Thumb	spring annual	seed	n.d.	spring	excellent	good	crops, gardens gullies	sow clean seed
Rumex sp. Dock, Sorrel	spring annual	seed	60,000	spring	excellent	decades	disturbed places, crops	prevent seeding

Sources: Georgia 1938; Gross 1924; Holm *et al.*

remain viable for decades (Gross 1924). Plants establish quickly from seed; each plant has a life span of three to five years. The seed is relatively heavy, so it does not travel far from the parent plant. Livestock eat the foliage from curled dock (*Rumex crispis* L.) and seeds germinate after passing through the gut, so animals provide one means of transportation. Dock is also a contaminant of seed grain, and thus contributes to crop contamination. Seeds from members of the dock family have high amino acid scores, with particularly high level of lysine (Table 12). Along with peas, lentils, and wild buckwheat, they complement cereals nicely.

Hairy tare (*Vicia hirsuita* (L.) S.F. Gray) is a slender annual found in hedges, fields, and grassy places.

Rarely Occurring Species

Twenty-two taxa occur only sporadically at Early Neolithic sites. Of these, thirteen occurred only once in the sites surveyed. The characteristics of selected species are summarized in Table 14.

Prehistoric Planting Strategies

The ecological indicators of the majority of the weedy species show a narrow range of variation (Table 15). Light requirements vary from partial shade to full sunlight; soil moisture varies from dry to moist, with only one species tolerating damp conditions; and pH varies from weakly basic to weakly acidic. Only slightly greater variation occurs in the nitrogen requirements. Most species indicate a rich to very rich soil, although the weedy legume tolerates a soil deficient in nitrogen, and both cleavers and rye grass indicate an intermediate soil fertility. The ecological data suggest crops were not grown on soils with deficient fertility. In Early Neolithic cultivation, fertilization may been practiced, harvest yields may have been stabilized below 1000 kg/ha, or the fields may have been fallowed. The first two possibilities imply that long-term or permanent cultivation of plots may have occurred. If that were the case, one would expect to see the establishment of weed communities adapted to cereal fields. Willerding (1979, 1980) and Groenman-van Waateringe (1971) argue there is no evidence for such weed communities until the Iron Age. This suggests fallowing was a component of Neolithic cultivation strategies.

Table 14

Characteristics of Rarely Occurring Weeds

Species	Growth Habit	Propagation	Seeds per plant	Seedlings Appear	Survival in Livestock Digestive Tract	Long-term Viability in Soil	Places in Which Weed Appears	Means of Control
Papaver dubium Long headed poppy	spring annual	seed	18,000	May	n.d.	n.d.	Cereals, pathways	hoeing, pulling
Polygonum lapathifolium Pale persicaria	annual	seed	19,300	spring	excellent	good	damp meadows	hoeing, pulling
Rumex crispus Curled dock	perennial	seed, upper taproot	100–60,000	fall, summer	excellent	decades	crops pastures	prevent seeding
Rumex acetosella Sheep's sorrel	perennial	seed, taproot	2,100	fall, summer	excellent	decades	cereals, meadows, hedges, woods	prevent seeding
Setaria viridis Green foxtail	annual	seed	n.d.	May-June	n.d.	decades	crops	hoeing, pulling
Solanum nigrum Black nightshade	annual & biennial	seed	178,000	spring	n.d.	decades	gardens, forest	hoeing, pulling

Sources: Georgia 1938; Gross 1924; Holm *et al.* 1977;

Table 15

Ecological Requirements of Key Weedy Species

Species	Light	Moisture	pH	Nitrogen
Chenopodium album Lamb's quarters	x	Moist	x	Rich
Chenopodium glaucum Glaucous chenopodium	Full light	Moist	x	Very Rich
Bromus secalinus Rye grass	Open light	x	x	x
Bromus sterilis Sterile brome	Open light	Moist	x	Inter-mediate
Echinochloa crus-galli Cockspur	Light	Moist	x	Rich
Galium spurium False cleavers	Light	Moist	Weakly basic	Inter-mediate
Polygonum persicaria Lady's thumb	Light	Dry	x	Rich
Vicia hirsuta Hairy tare	Light	x	x	Poor
Rumex sanguineus Red veined dock	Partial shade	Very damp	Neutral	Rich
Solanum nigrum Black nightshade	Partial shade	Moist	Neutral	Rich
Lapsana communis Nipplewort	Half Shade	Moist	x	Rich
Setaria viridis Green foxtail	Half shade	Dry to Moist	x	Rich

Source: Ellenberg 1979 x= Highly variable

Groenman-van Waateringe (1971) evaluated the repeated co-occurrence of weedy species at Early Neolithic sites in view of their phytosociological associations. He argues that nipplewort, brome grass, and red veined dock are characteristic of thorny hedges surrounding small forest openings. He notes that these particular thorny hedges are not browsed by livestock and suggests they would have made an effective barrier to livestock—both by excluding them from ripening crops, and by corralling them while feeding on crop stubble. Groenman-van Waateringe suggests that the fields were small and surrounded by thorny hedges. Willerding (1980) concurs.

There are three ramifications for planting strategies. First, if the plots were small, they probably represented scattered, individual household fields. This conforms to a general model of household level production found throughout horticultural societies. Second, the plots would have been subjected individually to mouse predation, rusts and blights, and, if widely scattered, perhaps to the localized effects of severe storms. Experiments in the Rhine valley (Lüning and Meurers-Balke 1980) suggest that small, isolated plots are particularly susceptible to mouse predation and that mice can be responsible for losses of up to 92% of the crop. Third, approximately six years are needed for an impenetrable thicket of the type Groenman-van Waateringe suggests to develop. This implies that fields may have been maintained through a ley fallow.

Finally, the presence of fall and spring germinating weeds, particularly the high frequencies of rye grass, suggests planting may have occurred both in the fall and the spring. As discussed earlier, both Willerding (1980) and Groenman-van Waateringe and Groenman-van Waateringe (1979) argue that weed communities in cereal fields did not develop until the Iron Age. If planting occurred in the fall, then the labor requirements for planting could have been divided between the fall and the spring; however, fall is also a time of increased labor for crop processing and preparing for winter. Finally, crop yields from winter wheats are substantially larger than those from spring wheats. Spring planting, on the other hand, would require that all field preparation be done in the late winter/early spring, as soon as the ground was sufficiently dry to work. In years with a late or wet spring, planting would have been delayed. Such a delay could have resulted in both lowered wheat yields and further delay in planting peas, lentils, flax, and poppies. The effects of both a mixed fall-spring planting strategy and a pure spring planting will be considered in the dynamic model presented in Chapter 6.

Summary and Conclusions

Early Neolithic cultivators planted cereals and legumes, as well as crops rich in vegetable oils. General climatic conditions would have influenced the basic crop yield. Yields could have been further affected by mouse and bird predation, crop diseases, severe storms, and the skills of an individual farmer. Analysis of the ecological requirements of weed assemblages found repeatedly with charred remains of domestic crops suggest that Early Neolithic cultivation strategies included small, scattered fields of wheat. Poppies, peas, lentils, and flax, on the other hand, were probably planted in relatively small garden plots. Cereals could have been planted either in the fall or spring, but the garden crops would have to have been planted in the early spring.

If cereal plots were small, cultivation would probably have been organized on a household level. The weed assemblage suggests that all plots were located on fertile, well-drained soil. Harvest yields could be expected to be uniform, but if the plots were isolated, they would have been subjected individually to rusts and blights, bird and mouse predation, or severe storms.

5

Neolithic Subsistence II: Livestock

This chapter examines the influence that livestock had both on structuring Early Neolithic subsistence strategies and changing the deciduous forest environment. Pertinent aspects of each animal's life cycle, feeding ecology, and nutrition will be examined. A series of reference herds and flocks of varying sizes will be established and their annual expected meat offtakes and milk yields estimated. Finally, grazing, browsing, and fodder requirements are suggested for each herd and flock. The estimates will be used in Chapter 6 to construct a dynamic model of Early Neolithic subsistence strategies.

Neolithic farmers faced one major constraint that affected stockbreeding: livestock needed to be housed through a four-month winter. Pollen evidence suggests specialized communities of pastures and hay meadows did not develop until the Bronze Age, but, until recently, the lack of specialized pastures and meadows did not seem to be a problem. Arguments that loess soils were sparsely forested (Gradmann 1902) were widely accepted and allowed archaeologists to assume that expanses of lightly forested grasslands existed. In a sparsely forested environment, stockbreeding would have been a relatively successful venture. But it now appears loess soils were as heavily forested as were soils of other types in Central Europe, and that Early Neolithic communities were located primarily in forested environments.

This has two ramifications for subsistence farming. First, cultivated fields would have been located in forest clearings. The tendency for forest succession to occur would have been a constant problem, and forest herbs and woody plants would have invaded cleared fields regularly. Second, Early Neolithic livestock would have had to survive primarily on forest browse with grazing on natural meadows, crop stubble, and fallow plots supplementing their spring-summer-fall browsing. The selective grazing and browsing patterns of livestock, particularly goats (who thrive on woody plants), may have played a significant role in maintaining open fields. In addition to providing critical fats and proteins for the Neolithic diet, the ecological effects of grazing and browsing may have been crucial for successful deciduous forest cultivation.

Ruminants

Ruminants transform indigestible plants into high-quality foods for the human diet, and they provide essential raw materials for clothing and tools. Herd size and composition need to be reconstructed in order to determine the requisite grazing, browsing, and pasture lands as well as to estimate the amount of meat and milk that would have been available on an annual basis. A logical circularity arises when one tries to reconstruct herd sizes, for decisions on herd size undoubtedly rested on the amount of available pasturage and browse at each village.

One method for resolving this circularity is to establish reference herds and flocks of arbitrary sizes. These figures can then be used to estimate the quantities of meat and milk that each herd would have produced and to project the area needed for grazing and browsing. The meat and milk yields and graze/browse/fodder requirements can then be used to establish a series of stockbreeding models. Finally, the stockbreeding parameters can be combined with cultivation parameters to identify a range of potential subsistence strategies for Early Neolithic villagers. Three variables are significant in examining Neolithic stockbreeding: (1) yearly meat offtake, (2) annual milk yields, and (3) the grazing, browsing, and fodder requirements for each herd. Before discussing cattle, sheep, and goats individually, the basic assumptions concerning the estimation of these three parameters will be introduced.

Meat Offtake

In order to derive estimates of meat offtake, three assumptions have been made. First, herds and flocks are maintained at specific sizes, and there is no herd growth. The lambs, kids, and calves in effect are assumed to be replacements for animals that are slaughtered or that die from natural causes during the year.

Second, animals that appear to be ill are slaughtered prior to their death, and the meat is eaten (Campbell 1964; Cranstone 1969; Dyson-Hudson and Dyson-Hudson 1970; Evans-Pritchard 1940; Lockwood 1975; Redding 1981). Livestock that die prior to being slaughtered can also be butchered and consumed, but their carcasses may not always be found. For example, an animal may have gotten lost, it may have fallen prey to wolves, or its carcass may have been eaten by either domestic or wild pigs (Grigson 1982b). Redding estimates that because of these factors, only 60% of available meat is actually consumed by the villagers. His figure is used. Third, a 50:50 sex ratio is assumed at birth.

Annual Milk Production

Milk is a superior food. It is rich in proteins, calcium, fats, and nutrients; furthermore, it is a renewable resource. Archaeologists commonly assume domestic animals were kept primarily as a source of meat rather than for their milk products. Although evidence of milk use from the archaeological record is sparse, circumstantial evidence for its use does exist. Milk in liquid form could have been held in skins or ceramic vessels. Skins would not be preserved in most archaeological contexts, so they cannot be expected to be recovered. Milk containers, as a class, may have differed stylistically from other ceramic vessels, but unless chemical tests are undertaken to identify vessel contents, it is unlikely milk vessels would be recognized. Ceramic strainers do appear in late Early Neolithic contexts. These have been interpreted as cheese strainers (Bogucki 1984; Milisauskas 1978), and it is not unreasonable to assume milk was consumed prior to the development of a cheese-making technology. Lacking a storage technology, Early Neolithic populations would have been limited in the amount of milk they could have stored, and production would have been for immediate consumption. Because milk provides such a superior food, the milk productivity of cattle, goat, and sheep herds must be examined.

Browsing and Grazing

Food selection among ruminants varies from animal to animal and season to season, as well as from species to species (van Dyne *et al.* 1980). Lacking upper incisors, cattle use their tongues to pin herbage to their lower teeth before tearing it off the plant. Because of this, cattle must eat plants that can be easily torn, and the presence of large quantities of woody plants in their pasture interferes with grazing, even though shrubs comprise up to 15% of their annual diet. Sheep have a cleft upper lip that allows them to graze much closer to the ground than cattle. They prefer to have more herbs in their diet than do cattle, but woody plants usually do not constitute more than 20% of their diet.

Goats are more cosmopolitan in their eating preferences than are either cattle or sheep, and "there is little in the way of green vegetation that a goat will not eat" (Mackenzie 1980:135). They seem impervious to thorns, and among their favorite weeds are thistles, stinging nettle, cleavers, and dock. Their favorite shrubs include brambles, briars, ivy, and heather (Mackenzie 1980:135). Goats have the ability to eat large quantities of coarse, relatively poor-quality fodder and to extract the best nutrients (see Redding 1981:53–80). To subsist on lower-quality fodder, goats must eat roughly twice as much as sheep, who require high-quality fodder (Mackenzie 1980:22). On good pastures, however, goats eat roughly the same amount as sheep. Goats browse preferably on woody plants, which constitute 70% of their diet, and they are important in clearing underbrush. Goats compete directly with deer for browse (Huston 1978:990).

During the Early Neolithic, Central Europe was covered by a thick deciduous forest. Pastures, as known today, are not a natural botanical community. Paleoethnobotanical evidence suggests that pastures had developed by the Bronze Age (Willerding 1979), but they were not in existence prior to the arrival of Early Neolithic populations. Until pastures developed, Neolithic livestock had to survive in a forested environment. Cattle and ovicaprids have complementary grazing and browsing patterns, and the combination of the three species would have made an effective use of available forest cover.

Cattle, sheep, and goats have different life cycles, fodder requirements, reproductive potentials, and both meat and milk production. The population ecology of each species is discussed separately below. Estimates of their grazing and fodder requirements as well as their meat and milk production are developed for use in the model Neolithic subsistence strategy.

Cattle

Cattle are prized as a source of meat, milk, blood, leather, and bone as well as acting as beasts of burden. Since cattle do not have a specific breeding season, calving can occur in any season. Weaning takes place after approximately 200 days, and calves mature over the subsequent two years. The first parturition for the majority of heifers occurs during their third year (Neumann and Snapp 1969; Perry 1984), although occasionally a heifer first calves in her second year.

Herd Composition

Distinguishing between the faunal remains of male, female, and castrated cattle can be difficult, and a satisfactory method has yet to be developed (Grigson 1982b:7). Müller (1964) identified bones from castrated cattle at an Early Neolithic site in Central Germany, and Bogucki (1982:109) argues that such bones are present at an Early Neolithic site in the Polish lowlands. Castration speeds weight gain and makes the animal easier to handle. Bogucki, on the basis of ethnographic data, suggests cattle were raised for meat as well as milk production, with minimum herd sizes were between 30 and 50 head.

Although cattle do not have a specific breeding season, a calving season can be created by allowing bulls access to cows only for a restricted period. There are particular advantages to a late winter/early spring calving season. First, the cows are stalled over the winter, so they can be closely watched during the later stages of gestation and can be helped in calving if necessary. Second, cows provide more and better milk on spring and summer pastures than they do on fall pastures and winter fodder. Third, spring calves are weaned by the onset of winter and their body weights are higher; thus they are prepared to withstand the winter. Grigson (1981) notes that there is very little outbreeding toward wild strains, and she suggests that Neolithic cows were intentionally bred with domestic bulls. Wild cattle roaming the Central European forests would have had access to the cows throughout the summer. Stockbreeders could have ensured that domestic bulls serviced the cows by breeding the cows in the spring. This would have resulted in a late winter calving, and it would have prevented potentially undesirable, wild traits from being introduced into the domestic herd. For the purpose of this analysis, it will be assumed that calving occurred in the spring.

Table 16

Reference Cattle Herd Compositions: Fall and Winter

	Herd Size		
	30-Head	40-Head	50-Head
Calves	0	0	0
Yearlings	7	10	12
Heifers	8	8	10
Steer	2	4	7
Cows	11	16	19
Bulls	2	2	2

This results in a set of calves that are roughly the same age, and in a lactation period that begins and ends within a few weeks for all cows.

Using Bogucki's minimum herd sizes and the assumption of a late winter/early spring calving season, the composition of Early Neolithic cattle herds can be estimated, if the birth and mortality rates of cattle are considered. Studies of unimproved cattle suggest that heifers first calve when they are 3–1/2 to 4 years old (Dyson-Hudson and Dyson-Hudson 1970:11). Contemporary studies of beef cattle indicate that 80% of the mature cows calve and, of the calves born, 20% do not survive to weaning (Perry 1984). Reference herd compositions can be suggested using these parameters and assuming that steers are slaughtered at 36 months (Bogucki 1982). The first represents the fall and winter herd composition (Table 16) these figures will be used to estimate winter fodder requirements. Table 17 shows herd composition one day after the hypothetical day on which all the cows calved. These herds represent hypothetical averages for estimating usable

Table 17

Reference Cattle Herd Compositions: Spring and Summer

	Herd Size		
	30-Head	40-Head	50-Head
Calves	9	13	15
Heifers	15	18	22
Steers	2	4	7
Cows	11	16	19
Bulls	2	2	2

meat and annual milk supplies, as well as spring-summer-fall browse and pasture requirements.

Estimated Annual Meat Offtake

Weight is a critical aspect in estimating the amount of meat that would be available from a heifer, steer, or cow. There is no general agreement on the weight of Neolithic cattle, although Müller (1964) suggests a withers height of 125–140 cm for mature females and 135–160 for adult males. Clason (1972) suggests a figure of 700 kg for mature males, and this figure has been adopted by several other archaeozoologists. More recently Wijngaarden-Bakker (cited in Bakels 1982) has suggested a weight of 400 kg for mature males from an early Bronze Age site. However, Grigson (1981) notes that Bronze Age cattle were smaller than their Neolithic ancestors. The estimate of 700 kg has been adopted for bulls, but it was lowered to 600 kg for steers, 550 kg for cows, 400 kg for heifers, and 75 kg for calves. The amount of usable meat is estimated at 50% of the live weight. Using these parameters, the annual meat offtake in kilograms can be determined (Table 18).

Table 18

Expected Annual Meat Offtake: Cattle

Age Group	30-Head		40-Head		50-Head	
	No.	Usable Meat (kg)	No.	Usable Meat (kg)	No.	Usable Meat (kg)
Calves	2	75.00	3	112.50	3	112.50
Heifers	2	400.00	4	800.00	4	800.00
Steers	2	600.00	3	900.00	4	1200.00
Cows	3	825.00	3	825.00	4	1100.00
Total	9	1900.00	13	2637.50	15	3212.50
Net meat offtake @ 60% consumption		1140.00		1582.50		1927.50

Estimated Annual Milk Production

Present-day beef industry sources suggest cattle lactate for 180 days (Neumann and Snapp 1969), but Dyson-Hudson and Dyson-Hudson (1970) note that in unimproved breeds in Ugandan herds, lactation lasted approximately 230 days. They also suggest that at best slightly less than half of cows with a calf produce milk beyond the needs of its calf, with an average daily surplus of 1.78 liters per cow. Milk has a density of 1.03 (OECD 1984), so a surplus of 1.8 kg is produced per cow. Assuming that 50% of the cows produce surplus milk, a lactation length of 200 days and Dyson-Hudson and Dyson-Hudson's figures, the annual surplus milk production for each herd has been calculated—see Table 19.

Browsing, Grazing, and Fodder Requirements

There is no agreement on the area a herd would need for browsing and grazing. Bogucki (1982:106) suggests that if cattle were grazed only in a deciduous forest environment, each mature

Table 19

Expected Annual Cattle Milk Yield

	30-Head	40-Head	50-Head
Total cows in herd	11	16	19
Cows with calves	9	13	15
Cows producing surplus milk	4.50	6.50	7.50
Daily milk surplus (kg)	8.10	11.89	13.72
Annual milk supply (kg)	1620.00	2378.00	2744.00

animal would require an average of one hectare per month. Bakels (1982) on the other hand, argues that had Early Neolithic cattle been grazed solely on pasture, they would have required approximately 1.5 ha per mature animal. Assuming that livestock derived 75% of their subsistence from forest browse and 25% from natural pastures as well as from cereal fields early in the growing season, crop stubble after harvest, and fallow fields, it is possible to estimate the amount of forest and pasture lands each reference herd would have required. Mature cattle would have required 6.0 ha of forest and 0.375 ha of pasture land. Estimating that one calf requires the equivalent of 15% an adult's ration (see Netting 1981:39), and that immature cattle require an average of 80% of a mature animal's portion, forest and pasture requirements were established. These are summarized in Table 20.

Cattle would have had to have been sheltered and fed for four months in the winter. While straw from the cereal crop could have been used as fodder, it should provide no more than 40% of the diet (NRC 1984). Estimates of the daily straw requirement are shown in Table 21. Annually, a 30-head herd would require

Table 20

Cattle Browse and Pasture Requirements

	Herd Size		
	30-Head	40-Head	50-Head
Forest (km^2)	1.70	2.29	2.97
Pasture (ha)	10.71	16.57	17.92

17,036.40 kg, a 40-head herd 22,887.60 kg, and a 50-head herd 28,684.80 kg. Assuming an average straw yield of 2200 kg/ha, a 30-head herd would require the straw produced from a 7.74 ha cereal crop, a 40-head herd would require that from 10.40 ha, and a 50-head herd, the straw from 13.06 ha.

Cattle would have to derive the remaining 60% of their winter fodder from hay or dried leaves. Mature cows in a contemporary Alpine village consume approximately 400 kg hay per animal per month, one-year-olds need 171 kg, and three-year-olds require 286 kg (Netting 1981:39). Using these figures and assuming yearlings consume the equivalent of 25% of a mature cow's requirements, the needed hay can be estimated for the three reference herds.

Difficulties arise in trying to estimate the number of hectares Early Neolithic villagers would have had to mow to collect the requisite hay. Modern hay meadows are well cared for to encourage a luxurious growth, and the best of these regularly produce over 4000 kg/ha (Ellenberg 1952). At the other end of the scale are natural meadows found on low-lying damp soils fringing marshes and bordering slow-moving streams. These provide the equivalent of 1470 kg/ha (Ellenberg 1952). Assuming shorelines provided one of the few natural meadows in the Early Neolithic, the 1470 kg/ha figure best represents the yields that could have been obtained from meadows in the Early Neolithic. This figure has been used to estimate the number of hectares required in natural meadows for hay (Table 22).

Table 21

Estimated Winter Cattle Fodder: Straw
(kg)

Age Class	Monthly Requirement	Winter Herd Total		
		30 Head	40 Head	50 Head
Yearlings	45.00	1260	1800	2160
Heifers	129.00	4128	4128	5160
Steers	175.00	1400	2800	4900
Cows†	196.75	8657	12592	14953
Bulls	196.50	1572	1572	1572
Total		17017	22892	28745
Ha		7.74	10.40	13.06

Source: NRC 1984 †Includes gestating cows

Cattle Requirements and Resources

Table 23 summarizes the estimated meat and milk yields as well as the estimated browsing, grazing, and pasture lands for each of the three reference herds. In the next chapter, an arbitrary village size of 34 individuals is used to develop an optimal farming strategy. Assuming a village of 34 inhabitants, the herds would provide between 33.5 and 56.7 kg of meat and from 47.6 to 80.7 liters of milk per person per year. These herds would require between 1.70 and 2.97 km^2 of forested land, 10.71 to 17.92 ha of pasturage, and 13.92 to 23.97 ha of meadow land in addition to straw from the cereal crop. Furthermore, full-time herders, probably older children, would be needed to tend the herd throughout the grazing season.

Table 22

Estimated Winter Cattle Fodder: Hay
(kg)

Age Class	Monthly Requirement	Winter Herd Total		
		30 Head	40 Head	50 Head
Yearlings	60.00	1680	2400	2880
Heifers	137.00	4384	4384	5480
Steers	240.00	1920	3840	6720
Cows†	240.00	10560	15360	18240
Bulls	240.00	1920	1920	1920
Total		20464	27904	35240
Ha		13.92	18.98	23.97

Source: NRC 1984

†Includes gestating cows

Table 23

Cattle Summary: Meat and Milk Yields;
Browse, Graze and Fodder Requirements

Herd Size (head)	Usable Meat (kg)	Milk Supply (kg)	Forest Browse (km^2)	Pasture (ha)	Cereal Straw (ha)	Meadow Hay (ha)
30	1140.00	1620	1.70	10.71	7.74	13.92
40	1582.50	2378	2.29	16.57	10.40	18.98
50	1927.50	2744	2.97	17.92	13.06	23.97

Goats and Sheep

Because of the high frequencies of cattle bones in Early Neolithic faunal assemblages, the potential significance of goats and sheep in the diet of Early Neolithic populations in Central Europe has not been seriously evaluated. The most comprehensive study of prehistoric goat and sheep herding was prepared for Middle Eastern Bronze Age flocks (Redding 1981). Despite disclaimers that the findings of this study may not be applicable to prehistoric goat and sheep populations outside the Middle East, many of the parameters discussed for modern herds (Redding 1981:53–137) correspond with European and North American findings (Mackenzie 1980; NRC 1981). In addition, Redding's conclusions are based on the evaluation of unimproved herds. This aspect of Redding's study is particularly valuable in developing an understanding of prehistoric ovicaprid populations. Until parameters for goat and sheep populations of the Early Neolithic period in Central Europe are established, those suggested by Redding provide a starting point for the analysis.

Life History

With a few differences relating primarily to the number of offspring per pregnancy and the ability to survive on low-quality forage, sheep and goats have a similar life history. Breeding is controlled by photoperiodicity, with the shortening of days triggering oestrus (Redding 1981:53). Since the magnitude of change in day length is affected by latitude, sheep and goats in northern climates have a more marked breeding season than sheep and goats near the equator. In temperate regions the breeding season for ovicaprids occurs primarily in September/October, with a few animals breeding slightly earlier or slightly later (Mackenzie 1980:217).

The age of first parturition in sheep and goats can be as early as one year, but it normally occurs at two years. Redding (1981:63) suggests that 12–15% of the female lambs and 30% of the female kids breed when they are six months old and bear their first offspring when they are one year old. Does and ewes can be expected to bear young for up to eight years. Gestation lasts five months, with lambing and kidding occurring in late winter, generally in February or March. Does frequently bear twins, while the majority of ewes have single births (Mackenzie 1980:25; Ryder 1983). Ethnographic sources from the Middle East suggest

Table 24

Ovicaprid Reference Herds: Winter Composition

Winter Flock Size (head)	Sheep				Goats			
	Lambs	Year-lings	Ewes	Rams	Kids	Year-lings	Does	Rams
15	0	3	10	2	0	4	9	2
20	0	4	14	2	0	5	13	2
25	0	5	18	2	0	6	17	2
30	0	6	22	2	0	7	21	2
35	0	7	26	2	0	8	25	2
40	0	8	30	2	0	10	28	2
45	0	9	34	2	0	11	32	2
50	0	10	38	2	0	12	36	2

the lambing rate is 0.80 for all ewes and the kidding rate is 1.20 for all does over the age of one year (Redding 1981). In unimproved varieties 61% of the ewes and does will produce milk (Redding 1981). The mortality rate for the first year of life is 32% for lambs and 45% for kids.

Flock and Herd Compositions

As with cattle, reference ovicaprid flock and herd sizes have arbitrarily been determined based on Redding's breeding and mortality rates (Tables 24 and 25). As with the cattle reference herds, the winter composition table will be used to estimate winter fodder requirements and the spring table will be used to determine spring, summer, and fall browsing and grazing requirements, as well as meat offtake and annual milk production.

Table 25

Ovicaprid Reference Herds: Spring Composition

Winter Flock Size (head)	Sheep				Goats			
	Lambs	Yearlings	Ewes	Rams	Kids	Yearlings	Does	Rams
15	8	3	10	2	12	3	10	2
20	12	3	15	2	16	3	15	2
25	14	4	19	2	22	4	19	2
30	19	5	23	2	26	6	22	2
35	22	6	27	2	31	7	26	2
40	25	7	31	2	36	9	29	2
45	28	8	35	2	41	9	34	2
50	31	9	39	2	46	10	38	2

Meat Yields

During the Early Neolithic, sheep had hair, but they had not yet developed woolly fleeces. Therefore, it is assumed that ovicaprids were kept primarily for milk. Two rams or bucks were needed for breeding purposes, but all other male offspring could have been slaughtered, if they did not die of natural causes. Of the female offspring, 32% of the lambs and 45% of the kids died prior to weaning. Yearlings consisted exclusively of immature females who survived their first year. These animals provided replacements for infertile does and ewes, or for animals that are killed by predators or that die from disease. As with cattle, 50% of the live weight would have been usable meat, and 60% of the total usable meat will be assumed to have been available for con-

Table 26

Expected Annual Meat Offtake: Sheep

Age Class	Flock Size (head)							
	15		20		25		30	
	No.	Kg	No.	Kg	No.	Kg	No.	Kg
Lambs	5	25.0	8	40.0	10	50.0	13	65.0
Ewes	3	37.5	4	50.0	5	62.5	6	75.0
Total		62.5		90.0		112.5		140.0
Meat @ 60% Use Rate		37.5		54.0		67.5		84.0

Table 27

Expected Annual Meat Offtake: Goats

Age Class	Herd Size (head)							
	15		20		25		30	
	No.	Kg	No.	Kg	No.	Kg	No.	Kg
Kids	8	40.0	14	70.0	16	80.0	20	100.0
Does	4	50.0	5	62.5	6	75.0	7	87.5
Total		90.0		132.5		155.0		187.5
Meat@ 60% Use Rate		54.0		79.5		93.0		112.5

Table 26 (Continued)

Age Class	Flock Size (head)							
	35		40		45		50	
	No.	Kg	No.	Kg	No.	Kg	No.	Kg
Lambs	15	75.0	17	85.0	19	95.0	21	105.0
Ewes	7	87.5	8	100.0	9	112.5	10	125.0
Total		162.5		185.0		207.5		230.0
Meat @60% Use Rate		97.5		111.0		124.5		138.0

Table 27 (Continued)

Age Class	Herd Size (head)							
	35		40		45		50	
	No.	Kg	No.	Kg	No.	Kg	No.	Kg
Kids	23	115.0	27	135.0	30	150.0	34	170.0
Does	9	112.5	10	125.0	11	137.5	13	162.5
Total		227.5		260.0		287.5		332.5
Meat@ 60% Use Rate		136.5		156.0		172.5		199.5

sumption. It is assumed that ewes and does had a live weight of 25 kg. Lambs and kids had an average live weight of 10 kg. The estimated annual meat offtake for sheep is presented in Table 26; the estimates for goats in Table 27.

Annual Milk Production

Lactation begins immediately at birth, and, for the first few weeks of their lives, lambs and kids subsist on milk. After this time they can begin to graze and browse. By the time lambs and kids start adding solid food to their diets, new spring growth has begun to appear. Throughout the spring and summer the amount of milk consumed decreases as grasses, forbs, and shrubs assume increasingly important roles in the diet. The quantity of milk produced daily and the length of lactation are important in determining annual milk production. Sheep and goats differ in both areas. Modern breeds of Central European sheep lactate for roughly five months, whereas modern goats can produce milk for up to ten months (Mackenzie 1980; Ryder 1983). Redding (1981) suggests the average length of lactation for unimproved breeds is 135 days for sheep, and 210 days for goats. His estimates will be used.

Contemporary Central European goats and sheep cannot be used to derive estimates of prehistoric milk yields because modern goat breeds throughout France, Switzerland, and Austria have been bred to be high-yielding milk producers. Their annual milk yields average more than 450 kg per doe, ten times the average of does in Greece, Yugoslavia, or Turkey (see OECD 1976, 1984 for annual country-by-country statistics). The best comparative data in Europe would probably come from herds in Greece or Yugoslavia. However, Greek does and ewes have shown a rapid improvement in milk yields over the past twenty years, with a 128% improvement occurring from 1962 to 1971 (OECD 1976, 1984). The yields of Yugoslavian ewes fluctuated only 3% during the same decade. Unfortunately, the Yugoslavian government prohibited goat keeping in the early 1950's, so there is no information on milk productivity in goat herds. Therefore, the Turkish herds provide the best available data.

Among Turkish does and ewes milk productivity fluctuated 1% in the decade from 1962 to 1971 (OECD 1976:134–135): annual goat milk production fluctuated between 70 and 71 liters, while sheep milk productivity varied from 46 to 48 liters. Dividing these figures by the average annual lactation of 135 days for sheep and 210 days for goats, the daily average yield is 0.28 liters

Table 28

Annual Sheep and Goat Milk Production

Winter Flock or Herd Size (head)	Sheep		Goats	
	No. Milkers	Kg Produced	No. Milkers	Kg Produced
15	6	170.10	6	340.20
20	9	255.15	9	510.30
25	12	340.20	12	680.40
30	14	396.90	15	850.50
35	17	481.95	17	963.90
40	19	538.65	20	1134.00
45	22	623.70	22	1247.40
50	24	680.40	25	1417.50

and 0.33 liters for sheep and goats respectively. These values are only 0.05 liters lower than the average daily yields suggested by Redding (1981). Because of the lack of adequate comparative data for contemporary unimproved Central European breeds, and because his values are close to contemporary production levels, Redding's daily production levels of 0.33 for sheep and 0.38 for goats will be used, but these must be adjusted for the slightly smaller size estimates for the Early Neolithic ovicaprids. Redding (1981: Table V-4) suggests a live weight of 40 kg for sheep and 35 kg for goats, whereas the estimated size of Early Neolithic sheep and goats is estimated to be 25 kg. Assuming that milk productivity varies in proportion to body weight, the milk production of Early Neolithic sheep would be 62% and goat milk production 71% of Redding's values. This would result in a daily production rate of 0.21 kg for sheep and 0.27 for goats. The expected annual milk

production for the reference sheep flocks and goat herds are presented in Table 28.

Graze, Browse, and Fodder Requirements

Sheep and goats of similar size require roughly the same amount of energy, but the protein requirements of goats are approximately 80% those of sheep of a comparable size (NRC 1975, 1981). Winter fodder estimates are based on energy needs in this model, so goats and sheep can be assumed to have the same gross requirements. The grazing, browsing, and fodder requirements for ten mature sheep and goats is equal to that of one cow (Mackenzie 1980; Netting 1981). Using the estimates presented in Table 23, and assuming that yearlings consume the equivalent of 80% and the lambs and kids approximately 15% of an adult ovicaprid portion, the estimated daily and annual straw requirements for ovicaprids are given in Table 29.

Meat and milk yields as well as browse, graze, and fodder requirements are summarized in Tables 30 and 31. Ovicaprid requirements are somewhat more complicated than those of cattle due primarily to the establishment of a larger number of reference flocks and herds. Meat offtakes vary from 37.5 to 138.00 kg for sheep and from 54.00 to 199.50 kg for goats. The differences between sheep and goats result from the greater birth ratio for goats in conjunction with a higher death rate among kids. Annual milk productivity also varies from 170.10 to 680.40 kg for sheep and from 340.20 to 1417.50 for goats. The difference between the two ovicaprids arises from a slightly higher daily yield and a longer lactation period for goats. As discussed in the next chapter, mutton and sheep's milk have a higher nutritional value than do goat's meat or milk. Thus, goat-keeping may not prove to be as advantageous as it appears.

Pigs

Early Neolithic pigs were small, averaging approximately 30 kg. Pigs can eat virtually anything. Their value to prehistoric communities is not just that they can convert inedible organic debris to nutritious proteins and fats. In consuming rotting vegetables, crop wastes, table scraps, and carrion as well as human and animal excrement, pigs provide some means of controlling village filth. That they convert this debris into meat is an added bonus.

Table 29

Ovicaprid Winter Straw Requirements

Winter Flock or Herd Size	Sheep			Goats		
	Daily (kg)	Annual (kg)	Annual (ha)	Daily (kg)	Annual (kg)	Annual (ha†)
15	9.08	1089.60	0.49	9.00	1080.00	0.49
20	12.67	1524.00	0.69	12.25	1470.00	0.67
25	15.57	1868.40	0.85	15.49	1858.80	0.84
30	18.47	2216.40	1.01	18.74	2248.80	1.02
35	21.71	2606.40	1.18	21.81	2596.80	1.18
40	24.50	2940.00	1.33	24.81	2977.20	1.35
45	27.86	3343.20	1.52	27.71	3325.20	1.51
50	30.77	3692.40	1.68	30.96	3715.20	1.69

Life Cycle

Pigs rut in late October to early November; gestation lasts four months, and sows farrow in the early spring. Lactation continues for two to four months—until mid-July. Litter sizes vary up to 13 piglets, but 5 or 6 are the norm. Sows usually farrow for the first time when they are one year old, and they can continue to breed for six years (Grigson 1982a:299). Pigs can live to be 20 years old. Among Neolithic assemblages, there appears to be little outbreeding to wild swine, thus suggesting sows were brought in for breeding in the late and autumn housed throughout the winter until they farrowed (Grigson 1982a). Litters may have been kept in the village until weaned in mid-summer, when the sows and their litters could have been put on crop stubble to eat the gleanings (Parsons 1962:229) or allowed to root in the forest. In late

Table 30

Sheep Summary: Meat and Milk Yields; Browse, Graze, and Fodder Requirements

	Estimated		Required			
Flock Size	Usable Meat (kg)	Milk Supply (kg)	Forest Browse (ha)	Pas-ture (ha)	Cereal Straw (ha)	Meadow Hay (ha)
15	37.50	170.10	9.72	0.58	0.49	0.26
20	54.00	255.15	13.08	0.79	0.69	0.34
25	67.50	340.20	17.52	0.98	0.85	0.43
30	84.00	396.90	19.17	1.18	1.01	0.52
35	97.50	481.95	22.98	1.39	1.18	0.61
40	111.00	538.65	26.25	1.58	1.33	0.69
45	124.50	623.70	29.52	1.78	1.52	0.77
50	138.00	680.40	32.70	1.97	1.68	0.86

Table 31

Goat Summary: Meat and Milk Yields; Browse, Graze, and Fodder Requirements

Flock Size	Estimated		Required			
	Usable Meat (kg)	Milk Supply (kg)	Forest Browse (ha)	Pas-turage (ha)	Cereal Straw (ha)	Meadow Hay (ha)
15	54.00	340.20	10.08	0.61	0.49	0.26
20	79.50	510.30	13.44	0.82	0.69	0.36
25	93.00	680.40	16.98	1.03	0.84	0.45
30	112.50	850.50	20.34	1.23	1.02	0.53
35	136.50	963.90	23.79	1.43	1.18	0.62
40	156.00	1134.00	27.24	1.57	1.35	0.68
45	172.50	1247.40	30.69	1.85	1.51	0.80
50	199.50	1417.50	34.14	2.05	1.69	0.89

September or early October, the pigs could have been allowed to fatten themselves on acorns prior to the fall slaughter.

Meat Production

Pigs are prolific, and they can be culled heavily, but they attain much of their body weight only after the second summer. Because of their prolific nature, pigs may have been an elastic resource: if fodder was low, they could have been slaughtered, but if resources were high, the pigs may have been fattened over the winter. Harcourt (cited in Grigson 1982a) suggests that immature pigs dominate some Neolithic faunal assemblages. If this is the case, then the majority of pigs may have been slaughtered prior to their first winter. A slaughter at the end of the summer would have provided a source of meat without the need to overwinter the animals, for only breeding sows would have been housed throughout the winter. This would have resulted in a low meat yield. Jacomet and Schibler (1985), however, note that the majority of pigs from a Middle Neolithic site in Switzerland were slaughtered at the age of 18 months. For this model, it will be assumed pigs were slaughtered at 18 months of age. As with the ruminants, a 60% use rate is estimated for meat consumption. The estimated weight for mature Neolithic pigs is 30 kg. Assuming 70% of the litter survives weaning, a total of 36 kg of meat and fat could be expected per litter each fall. Piglets that did not survive weaning would be butchered, but they could not be expected to provide more than two or three kg of meat, depending upon its age. Thus each litter could be expected to provide approximately 40 kg meat and fat, in addition to their services as village scavengers.

Browse and Pannage

If, as Grigson suggests (1982a), pigs were housed over the winter, they would have had to be fed. Table scraps, village debris, the remains of crop cleaning processes, and collected nuts would have been sufficient to carry yearlings through the winter. During the spring and summer, sows could have been allowed to root among ferns, returning to their litters at night. As mentioned earlier, crop gleanings and acorns may have provided the bulk of the late summer and fall diets. Both are high in carbohydrates and serve to fatten the animal. Acorns, in particular, are eaten with great enthusiasm, and 10 kg of acorns add 1 kg of weight on a pig. Acorns fall in the month of October, and they are grazed by wild swine, brown bears, badgers, squirrels, and birds as well as by domestic pigs. It is unlikely that acorns would be found on the forest floor throughout the winter; nonetheless, fern rhizomes could continue to be rooted out until the ground froze in December.

The success of pig husbandry depended to some extent on oak productivity. Oak saplings, however, are browsed by deer, goats, and sheep (Penistan 1974; Shaw 1974), and in areas with repeated grazing the future oak productivity can be seriously damaged. It takes oak trees 40 years to flower and produce their first acorns, but acorn production continues thereafter for 40 to 80 years before it begins to decline. Heavy grazing can have a long-term affect on the future acorn productivity of an area because grazing removes the seedlings and saplings that would provide forest renewal. This decline would influence not only domestic pigs, but also wild pigs, brown bears, badgers, and squirrels

Summary and Conclusions

Livestock provide a valuable source of domestic proteins, fats, and carbohydrates, but their impact on the environment can be dramatic. Even the smallest cattle herd would have required more than 1.7 km^2 of forest browse, and the combined ovicaprid herds could have added as much as 66.8 ha to this figure. The effects of ruminant browsing can be significant. When a forest is over browsed, herbs virtually disappear, saplings up to 4 meters high can be ridden down and defoliated, and trampling can result in unfavorable soil conditions for the trees that survive (Biswell and Hoover 1945). Most significant, woody vegetation is slow to recover. Overgrazing in just one year can result in a 50% decrease in available browse for the following year (Biswell and Hoover 1945). Thus the estimates for the Early Neolithic should be doubled to allow for at least a one-year hiatus for recovery between summers in which the area is used for browse.

The complementary grazing patterns of cattle, goats, and sheep can be beneficial. Cattle prefer grasses; sheep, herbs; and goats, woody vegetation. Grazing all three on fallow cereal plots would reduce the tendency for forest succession to occur. Ferns can also invade fallow cereal plots, and they can be difficult to eradicate. Pigs, however, relish ferns, and in rooting out the rhizomes pigs aerate the soil. Grazing livestock in the forest may ultimately change the forest composition. In grazing and browsing, livestock effectively remove their preferred species from the forest, thereby allowing the less-favored species to grow with little competition. Moreover, trampling can push large seeds into the soil, which otherwise would remain on the surface and fail to germinate. Finally, livestock can have a detrimental effect on wild game. In particular, ruminants compete directly with red deer, and red deer can survive only with little or no competition from livestock (McMahan 1968). Thus red deer probably cannot be expected to have been found in areas in which livestock browsed. For even the smallest cattle herds, this means that relatively few red deer might have been found within a 3.8 km^2 area of the forest near the village.

More precise estimates of the number of livestock kept at the village and of their impact on the landscape can be made once a general model of Early Neolithic subsistence is developed. This is done in the next chapter.

6
Optimal Farming Strategies

From the perspective of mutualistic interactions, archaeologists may have been right in suggesting the farmers could have provided grain to the foragers in exchange for resources or services. Ecological literature suggests that for long-term mutualistic interaction to occur, the resource of benefit provided by one population must cost little to produce but return disproportionately high benefits. Harvest fluctuations virtually guarantee that crop surpluses will occur, although perhaps at unpredictable intervals. This chapter explores constraints surrounding the Neolithic farming economy by determining the degree to which domestic foods could have fulfilled villagers' nutritional needs, the amount of land required, and the labor needed to produce the requisite foods. As mentioned earlier, dendrochronological studies cannot yet be used to correlate climatic fluctuations and crop yields; thus, at this time, the only way to investigate variability in Neolithic crop yields is by modeling harvest fluctuations. This chapter uses the data presented in Chapters 4 and 5 to dynamically model a Neolithic farming economy and to develop an optimal farming strategy.

Several issues complicate the creation of such a model. First, the crops provided not only wheat for the human population but also straw for fodder. Therefore, when estimating production levels, the nutritional requirements of both human and livestock populations must be considered. To do this, population levels must be established. But there are no hard-and-fast guidelines for determining either village population or the size of livestock herds. Furthermore, the significance of wheat in the Neolithic diet has not been established. In order to model optimal farming strategies, the issues of population size and the significance of domestic foods in the Neolithic diet must be resolved.

The variables are tightly interrelated. Harvest yields result from the interaction between the amount of land planted and the planting strategies; however, the population simultaneously determines the amount of wheat needed and provides the work force to plant and harvest the crop. By varying both the amount of cultivated land and the planting strategies within limits of available labor resources, harvests of different sizes are obtained. Thus, given village population and harvest yields, it is possible to determine the proportion of the annual diet that would be fulfilled by harvests from each combination of cultivated land and planting strategies. Then knowing nutritional requirements of cattle, sheep, and goats, the number of days winter fodder is needed, and the straw yield of the harvest, it is possible to estimate the size of herds that could be maintained through the winter. Given these herd sizes, the amount of available domestic meat and milk can be estimated.

By making calculations for several years, it is possible to identify optimal combinations of the amount of wheat in the diet, per-capita allotment of cultivated land, planting strategies, and herd sizes. The simulation developed here estimates these variables for a run of 100 years for a given amount of cultivated land per-capita in conjunction with specific planting strategies, but holding village population constant. From the results of the simulation it becomes possible to identify an optimal level of wheat in the diet, the number of livestock that could be maintained, and the amount of food livestock contributed to the diet. The specific parameters and assumptions shaping the simulation are introduced before developing an optimal farming strategy. The code for the simulation will be published elsewhere.

Modeling Neolithic Food Production

The flow diagram (Fig. 2) shows the organization of the simulation. The program first generates a reference village composed of individual households, establishes the demographic profile of each household, and estimates the daily and annual caloric requirements for each. The annual harvest is determined for each household, and the requisite amount of wheat to meet the household's annual needs is deducted from the harvest. One stipulation of the program (discussed below) is that each household maintain a year's supply of wheat in reserve. If the

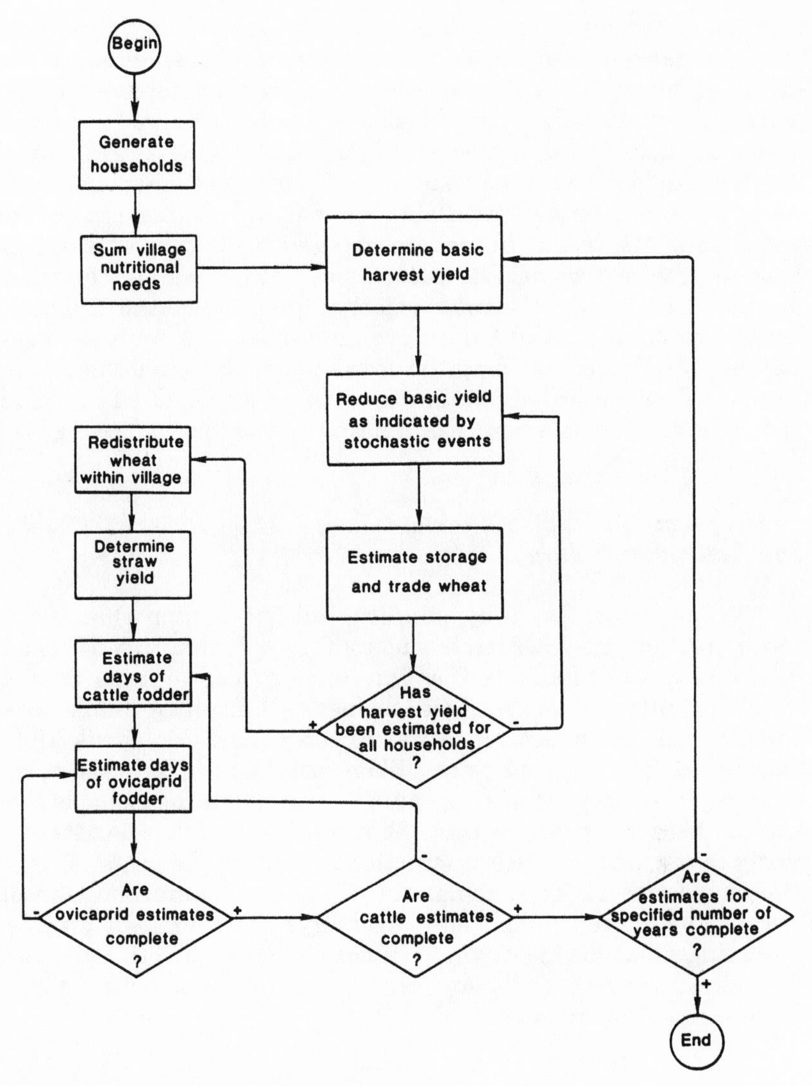

Figure 2. Simulation Flow Diagram

harvest is not sufficient to cover the household's requirements, then the deficit is subtracted from the household's stores. When sufficient wheat is not available in the household stores, then it is "borrowed" from other households. If each household's harvest fulfills its annual requirements, then surplus wheat is added to the household's emergency stores. Wheat remaining after the stores are replenished is available for lending to other households or for trade. Once the annual storage and trade surpluses are estimated, the straw yield is calculated. This yield is compared against the ruminants' winter fodder requirements and the number of available days of fodder are determined. After the fodder has been estimated the program loops to calculate another year's harvest. The program terminates after running for 100 years. The following discussion examines key factors affecting the simulation.

The Reference Village

The model is for a hypothetical village rather than for a known settlement. Research suggests Bandkeramik settlements varied from small hamlets of only a house or two to large villages of up to 50 houses, but archaeologists currently think most Bandkeramik settlements had five to ten households (Kruk 1980; Lüning 1982). The reference village established for this model consists of six households. A village size of six households was selected because it falls within the range most archaeologists currently favor, and because interactions between six units is, from the perspective of group dynamics, an optimal number (Johnson 1982). The figure is admittedly arbitrary; nonetheless, it provides a starting point for identifying critical variables affecting the subsistence strategies. Once key parameters are established, larger or smaller villages can be used.

Generating the Households and Estimating Their Nutritional Needs

The second step in the simulation was to determine the demographic profile of each household in order to estimate the annual amount of food required per household. The size of each household was generated randomly, and each household had an equal chance of a family having 4, 5, 6, 7, or 8 members. Food requirements vary according to an individual's age and sex. Thus, having obtained the size of each household in the village, it was necessary to determine the number of adults and children. The average life expectancy in the Central European Neolithic lay between 20 and 25 years (Scheffrahn 1969), with a 50% mortality rate before adulthood. Weiss (1973) has established demographic tables for prehistoric populations. His mortality Table M:22.5–30.00 shows a life expectancy of 22.5 years, and it has a mortality rate of 48.1% by the age of 15 (individuals over the age of 15 are considered to be adults). Since these figures correspond to those suggested by Scheffrahn, Mortality Table M:22.5–30 was used to determine the demographic profile of each household in the village. This table suggests an adult:child ratio of 52:48. Using this ratio, the number of adults and children per household were calculated (Table 32).

The caloric and nutritional requirements of the adults were estimated using the nutritional requirements listed in Table 1. Children, however, require a disproportionately large quantity of food in relation to their body size. In order to determine the requirements of the children, it was necessary to estimate the number of children in age cohorts corresponding to those in Table 1. The number of children per household was low, and fractions were rounded to the nearest whole number; therefore, it proved impractical to set precise proportions for all the age groups in Table 1. Instead, four cohorts were created. The infant cohort included babies through 11 months. Young children aged 12 months to 4.99 years were placed in the second age group. Children from 5 years to 9.99 years were placed in the third cohort, and children from 10 years to 14.99 years were placed in an older child cohort.

Following proportions suggested by Weiss, 31% of the children were placed in the 1–4 year old and 31% were placed in the 5–9 year old age groups. The membership of the infant and the 10–14 age groups were low, but the 10–14 year old cohort was larger in the mortality table. At this point, the selection had to be manipulated, otherwise the remaining children would have always been placed in the same cohort, and this would have resulted in a skewing toward either the infant or the older child cohorts. To avoid skewing, 12.6% were put in the 10–14 year old cohort. After estimating the 1–4, 5–9, and 10–14 cohorts, the remaining child (if there was one) was randomly assigned to either the infant

Table 32

Reference Village Demographic Profile

Household	Size	Children					Adults		Total Daily Calories	Total Annual Calories
		Infants	1–4	5–9	10–14		Women	Men		
					Girls	Boys				
A.	4	0	1	1	0	0	1	1	8,570	3,167,950
B.	5	0	1	1	0	0	1	2	11,570	4,286,050
C.	5	0	1	1	0	0	1	2	11,570	4,286,050
D.	4	0	1	1	0	0	1	1	8,570	3,167,950
E.	8	0	1	1	1	1	2	2	18,940	6,992,900
F.	8	1	1	1	1	0	2	2	16,740	6,189,900
Total	34	1	6	6	2	1	8	10	75,960	28,090,800

or the 10–14 year old cohort. The older-child cohort was further divided into boys and girls assuming a 50:50 ratio.

The daily and annual caloric requirements were estimated for each cohort by adjusting information in Table 1. The 1–4 year-olds were allotted 1360 calories per day; the 5–9 year olds, 2010 calories; the 10–14 year old boys 2750; and the 10–14 year old girls 2420. It was assumed that infants would have nursed through their first year; therefore for each infant in a household, an additional 550 calories per day were added to one woman's caloric requirement. The annual caloric estimates include an allowance for six weeks of heavy labor for all adults at a daily rate of 400 additional calories for each woman and 550 additional calories for each man. The reference population, its demographic profile, and the daily and annual caloric requirements are presented in Table 32.

Population growth can be expected to have occurred throughout the Neolithic, and it might be argued that population dynamics should have been included in the simulation. There is one major problem in adding population dynamics to the model at this stage. Population dynamics are not simply a function of biology. Cultural rules affect the size and shape of potential marriage networks, and kinship rules determine suitable marriage partners (Wobst 1974). Attempts to maintain a viable population depend on an influx of suitable marriage partners into the village, as well as on the migration of surplus individuals from the village. A consideration of population dynamics must be made from a regional perspective and should incorporate several villages. Unfortunately, suitable settlement pattern data are not yet available for the Early Neolithic. Moreover, studies of the dynamics of Early Neolithic social organization are in their infancy, and the principles of Early Neolithic kinship and marriage networks are only vaguely understood at best. For these reasons, and because this examination emphasizes economic conditions of Mesolithic/Neolithic interaction, the population size and its demographic profile are held constant.

Crop Yields

Among tribal horticulturalists, production and consumption occur at the household level (Brookfield and Brown 1963; Carneiro 1961; Freeman 1970; Izikowitz 1979; Scudder 1962; Titiev 1944).

Available paleoethnobotanical data indicate that Early Neolithic fields were small and partially shaded, and this suggests the fields were scattered throughout the forest in small clearings. The small size is compatible with a household level of production and conforms to the general model of cultivation among tribal horticulturalists. Therefore, the simulation was oriented toward household-level production and consumption. General climatic conditions would have affected the entire region, although some variation could result from differences in the topographic orientation of a field. If the fields were small and scattered (as paleoethnobotanical data suggest), it is unlikely that every plot would follow the same development. Instead, each plot would have been subjected independently to random events, such as mouse or bird predation; destruction by loose livestock, wild pigs, or red deer; infestation by weeds, rusts, or blights; localized thunder storms; and differences in each household's experience and skills. Rather than estimate the crop for the village as a whole and dividing that among the six households, harvest yields were determined separately for each household[1].

The interaction of six major factors affected crop yields. The first is field size. As discussed above, one of the purposes in developing the simulation was to determine an optimal per-capita field allotment. The simulation was run for 100 years using arbitrary allotments of 0.35, 0.40, and 0.45 ha per person in three separate runs. The amount of land planted for each household was determined by multiplying family size by the per-capita allotment. Since infants received only milk, a field allowance was not made for any infants in a household.

The second major factor is whether the crop was a winter wheat or spring wheat. Winter wheats have a yield that can be up to 30% larger than spring wheats. If wheat is planted only in the fall, then immediately after one crop is harvested, labor is needed to cultivate the soil and plant the next crop. This demand for labor comes at a time when labor is also needed to prepare for the coming winter. Furthermore planting only winter wheat is a risky business, for historical records suggest there is a 33% chance that a winter crop will have to be resown in the spring. A mixed

---

[1]Hegmon's (1985) detailed examination of restricted, rather than complete sharing, corroborates my arbitrary decision for a household level production and consumption.

fall-spring planting strategy would minimize the risk by allowing cultivators to plant half of their crop in the winter and half in the spring. A mixed planting strategy would allow labor to be spread throughout the year rather than concentrated in either the fall or the spring. It would also require smaller field allowances because of the higher crop yields. The potentially higher yields, however, must be offset by the need to maintain sufficient grain to reseed the winter crop should it fail. The allocations of 0.35, 0.40, and 0.45 ha per person were run once using a spring-only planting strategy and once using a mixed fall-spring planting strategy. In all, the simulation was run six times.

Historical records also indicate that the species and variety of wheat can affect yields. In particular emmer-spelt has a slightly higher average yield than einkorn (Table 3), but its standard deviation is larger. Emmer is generally assumed to have been the most common wheat at Bandkeramik sites. However, a review of quantified data from sites in the Rhineland and in the Low Countries suggests the dominance of either species varies regionally (Bakels 1978, 1984; Bakels and Rousselle 1985; Knörzer 1971a, 1973, 1980). The proportion of emmer to einkorn wheat was set at 1:1, with neither dominating.

As discussed in Chapter 4, climate and weather conditions are the most important variables affecting harvest yield. A basic hectare yield was determined for each year the simulation was run. The basic yield for each year was determined by using the standard deviation of historic crop yields. The area under a normal distribution was divided into discrete classes of Z-scores from −3.00 to +3.00 at intervals of 0.25. The percentage of the area under a normal distribution that each Z-score covered was then determined. This formed a sequence of 22 discrete classes, each with an lower and upper limit falling between 0.00 and 1.00. A number between 0.00 and 1.00 was randomly generated to determine the Z-score used in estimating the basic yield. The standard deviation was multiplied by the Z-score to obtain a gross standard deviation, and this amount was added to or subtracted from the average yield to obtain the basic kg/ha yield.

The basic yield was established for each year the simulation was run. Once a year's basic yield was established, it was used for every household—but each household's "harvest" was independently subjected to a series of events that could further reduce crop yield. Three events were particularly important in reducing yield. Mouse predation was the most significant.

Steensberg (1979) and Lüning and Meurers-Balke (1980) have shown that wood mice can pose a serious threat to small fields grown in forest clearings. The mice in effect harvest an ear of wheat, remove the kernels from the ear, and either eat the kernels or store them in their burrow. Lüning and Meurers-Balke report that in experimental fields mice destroyed up to 92% of the crop, although the average destruction was 30% for fields with moderate weed infestation and 50% with few weeds. For the simulation, the maximum crop loss that could occur due to mice infestation was set at 30%.

Severe storms or predation by livestock and wild animals posed the next most common sources of crop loss. Marauding game or livestock would have also posed a severe threat while crops were ripening. An arbitrary 2% chance of losing respectively 30%, 50%, 70%, or 90% of the crop to either storm damage or marauding livestock and game was allocated. There was a further 1% chance of losing the entire crop to storm damage, livestock, or game.

The final factor taken into consideration was a collective set of events that can affect the crop: rusts and smuts, snail infestation, bird predation, the general ability of the household to adequately care for their fields—particularly to perform weeding at a regular interval—and storage losses. It was arbitrarily decided that the collective category could account for up to a 10% reduction in each household's harvest yield.

In running the simulation, the basic kg/ha yield was determined and this figure was multiplied by the household's field allocation to obtain its gross yield. The gross yield was then reduced by up to 30% if mouse predation was severe. The remaining crop was subsequently reduced by up to 100% due to storms and livestock or game predation, and if crops remained perhaps by a further 10% due to miscellaneous factors. After deducting the specified amounts, the net harvest yield was obtained for each household. The caloric value of this harvest was then determined by multiplying the net yield by 3300, the number of calories per kilogram of wheat (Watt and Merrill 1975).

The caloric value of each household's harvest was divided by the daily and by the annual caloric needs of the household. This determined the number of days the harvest would fulfill the household's caloric requirements and the kilograms of surplus or deficit wheat. As mentioned above, the *daily* caloric requirements do not include an allowance for heavy labor, but the *annual* requirements include an allowance for six weeks of heavy labor for all adults.

Wheat Storage

One purpose in the simulation was to identify the optimal level of wheat in the Early Neolithic diet. In order to do this, calculations were made according to whether wheat provided 100%, 85%, 80%, 75%, 65%, 60%, 55%, 50%, or 45% of the total daily and annual calories. A consistent feature of tribal cultivators is that they attempt to maintain a minimum of a one-year supply of their primary crop, if the crop is amenable to long-term storage. One specification of this model, therefore, was that each household maintain a one-year supply of wheat. A further stipulation was that each household must fill its wheat supply before its harvest surpluses could be used in trade. Obviously, a positive correlation exists between the amount of wheat in the diet and the amount of land that must be planted: the more wheat needed, the more land that must be planted. The amount of wheat in the diet also affects the required amounts of wheat storage. If a household keeps a one-year supply as emergency storage, then stores must be proportionately larger when the household depends on wheat to fulfill a large proportion of its diet. Additionally, the more a household depends on wheat for its primary source of calories, the more vulnerable it becomes to annual fluctuations in harvest yields.

If surpluses occurred after deducting the household's annual requirements, these were added to the wheat stores until the one-year supply was filled. Surpluses remaining after the household's stores were filled became available for trading. When deficits occurred, they were subtracted from the wheat stores of the household. If wheat stores of the household were exhausted, then the requisite amount was taken from the trade wheat of another household. If no trade wheat was available in the village, the needed amount was subtracted in equal portions from the wheat stores of the other households. When all household wheat was exhausted, a total deficit for the village was recorded as a famine year.

Straw Production

The total straw production was estimated by maintaining a running tally of the number of hectares "planted" for each household and multiplying this by a basic straw yield. Mice in effect harvest the wheat in order to obtain the kernels. Their activity would reduce the kernel yield although livestock could eat the straw when grazing on crop stubble. The effects of mouse predation were not taken into account when determining the village's net straw yield. Similarly, severe storms and miscellaneous detrimental events would have affected primarily the kernels, and not the culms and leaves. As with the kernels, the mean and standard deviation of historic einkorn and emmer-spelt straw yields were obtained from historical records (Table 3). Kernel and straw yields showed no correlation when the historical yield data were used; therefore, straw yields were determined independently from the kernel yields. The Z-scores were divided into 22 classes, and, as was done in the case of the kernels, a random number generator was called to determine the Z-score. The standard deviation was then multiplied by this figure, and the result was added to (or subtracted from) the mean.

Livestock Fodder

Straw is not a particularly nutritious fodder, but it provides bulk in the ruminant winter diet. As discussed in Chapter 5, straw should supply at most 40% of the daily fodder. Using tables provided by the National Research Council (NRC 1975, 1981, 1984), the daily energy requirements for the reference herds of cattle, sheep, and goats were established. Forty percent of the figure was used as the number of kilograms needed daily for each reference herd. The simulation estimated the number of days the straw would feed the ruminants by subtracting the needs first of the cattle herd, then of the goats, and finally of the sheep. The calculations were made first on the basis of a 30-head cattle herd. The remaining straw was evaluated to determine the number of days it would supply goat herds and sheep flocks, moving progressively from 15 to 50 head in increments of 5 for each reference goat herd and sheep flock. The calculations were repeated after subtracting requirements for the 40-head cattle herd, and finally they were repeated again after subtracting those for the 50-head cattle herd.

Simulation Results

The simulation was initially run using a spring planting strategy with per-capita field allotments of 0.35 ha, 0.40 ha, and 0.45 ha. It was run again with a mixed planting strategy for each of the three field allotments. Each of the six runs was for 100 years, and randomly generated numbers determined the outcome of stochastic events every year. In order to be able to compare the relative efficiencies of different combinations of per-capita field allotments and planting strategies, identical random seed numbers were used at the start of every run. This ensured that the same events occurred in the same year of every run. Variation could thus be attributed to the per-capita land allotment (when the planting strategy was held constant) or to a difference in planting season (when the land allotment was held constant), rather than to differences in stochastic events.

Each run was not simply the reiteration of the same conditions 100 times. A critical aspect of the program was that wheat surpluses and deficits from previous years affected the availability of village supplies and trade surpluses. A poor harvest in one year had to be covered from the available supplies. This in turn lowered household supplies, which had to be replenished in subsequent years before surplus wheat could be used for trade. Identifying an optimal planting strategy and per-capita ha allotment required evaluating the volume and predictability of trade surpluses and the frequency of famines.

Trade surpluses and famines occur at different levels of analysis. Trade surpluses occur primarily at the household level. At the end of a harvest, each household must have sufficient wheat to fulfill their annual nutritional needs plus a one-year emergency store of wheat. Any remaining wheat becomes available either for households within the village whose stores were depleted and who had not been able to meet their requisite annual needs or for trade external to the village. Table 33 shows the proportion of years in which at least one house had a trade surplus for each of the planting and field allotment strategies, with wheat providing from 45% to 100% of the diet at 5% increments. Famines, on the other hand, are village-wide phenomena.

Table 33

Frequencies of Trade Surpluses
(% of years)

Planting Strategy	% Wheat in the Diet									
	100	85	80	75	70	65	60	55	50	45
Spring										
0.35 ha	0	0	0	0	0	4	52	79	89	96
0.40 ha	0	0	0	3	41	72	87	93	98	99
0.45 ha	0	1	28	61	80	89	94	98	100	100
Mixed										
0.35 ha	0	0	0	5	17	42	51	73	82	86
0.40 ha	0	7	17	37	50	67	76	82	82	96
0.45 ha	2	33	48	61	73	80	84	87	96	100

When a household deficit occurs, wheat can be taken from trade surpluses of other houses. If trade surpluses are not available, wheat is taken from household stores. A famine occurs when all trade supplies and household stores have been exhausted, but the annual needs of all households in the settlement have not been met. Table 34 shows the proportion of famine years for each per-capita field allotment with wheat providing from 45% to 100% of the annual diet for both planting strategies. Specific factors affecting spring-planted crops will be discussed first.

Two factors combine to cause famines in villages with spring-planted crops. The basic crop yield is the more significant. The availability of net stores within the village is the second major factor. Household stores are depleted when a household has a poor harvest due to a low basic yield and random events that decrease the household's crop. The simulation showed that when moderately poor harvests occur for three out of four sequential years, or very poor harvests occur two out of three years, stores are depleted throughout the village. Households become incapable of providing the requisite grain, and once stores are reduced, any deficits result in settlement-wide famine. Random events affect-

Table 34

Famine Frequencies
(% of years)

Planting Strategy	% Wheat in the Diet 100	85	80	75	70	65	60	55	50	45
Spring										
0.35 ha	100	94	90	84	66	32	5	0	0	0
0.40 ha	98	82	66	42	8	0	0	0	0	0
0.45 ha	89	41	12	2	0	0	0	0	0	0
Mixed										
0.35 ha	94	76	63	49	28	14	9	0	0	0
0.40 ha	82	46	28	14	10	2	0	0	0	0
0.45 ha	57	15	11	7	0	0	0	0	0	0

ing individual households exacerbate bad situations. They are generally not significant in determining whether a village-wide famine occurs, but after a series of years with low harvests, they may indeed set off a famine.

Deficits in harvests of mixed fall-spring planting strategies are somewhat more complicated. First, a basic stipulation of the simulation is that each household must maintain a year's supply of wheat, including sufficient grain to reseed the fields if necessary. When the crop is "replanted," the requisite seed grain is deducted from the household's wheat stores. If a household does not have sufficient stores to replant the wheat, wheat is taken equally from all other households. Winter crops must be replanted one year in three, and this has three consequences. First, each household must store an additional 140 to 360 kg seed grain, depending on the number of hectares the household plants (Table 35). This raises the threshold that must be met before trade surpluses can be accumulated; but it simultaneously increases the storage buffer, for these stores can be used if deficits occur. Fall planting presents one critical drawback. If winter crops must be replanted several years in a row, and if the har-

Table 35

Minimum Wheat Stores

	Added Seed Grain (kg)	% of Wheat in the Diet									
		100%	85%	80%	75%	70%	65%	60%	55%	50%	45%
Households 1 and 4											
Spring Planting		959	815	767	720	672	627	576	528	480	432
Mixed Planting:											
0.35 ha	140	1100	955	907	860	812	764	716	668	620	572
0.40 ha	160	1120	975	927	880	832	784	736	688	640	592
0.45 ha	180	1139	995	947	900	851	804	756	708	660	612
Households 2 and 3											
Spring Planting		1298	1307	1230	1153	1076	999	923	846	685	584
Mixed Planting:											
0.35 ha	174	1472	1481	1404	1327	1250	1173	1097	1020	823	758
0.40 ha	200	1498	1507	1430	1353	1276	1199	1123	1046	849	784
0.45 ha	224	1521	1531	1454	1377	1300	1223	1147	1070	873	808
Household 5											
Spring Planting		2118	1800	1694	1588	1482	1377	1271	1165	1059	953
Mixed Planting:											
0.35 ha	280	2398	2080	1974	1868	1762	1657	1551	1445	1339	1233
0.40 ha	320	2438	2120	2014	1909	1802	1697	1591	1485	1379	1273
0.45 ha	360	2448	2160	2054	1949	1842	1737	1631	1525	1419	1313
Household 6											
Spring Planting		1875	1593	1500	1406	1312	1218	1125	1031	937	844
Mixed Planting:											
0.35 ha	244	2119	1837	1744	1650	1556	1462	1369	1275	1181	1088
0.40 ha	280	2155	1873	1780	1686	1592	1498	1405	1311	1217	1123
0.45 ha	314	2189	1907	1814	1720	1626	1532	1439	1345	1251	1158

vests are poor, households may enter a decreasing spiral. A point can be reached at which the stores do not hold sufficient grain to replant the crop. Seed must either be borrowed from neighboring villages (as was assumed for this simulation), or the amount of resown land reduced. The latter course of action decreases the grain harvest, and it also affects the livestock, for the hectare reduction results in a lowered straw production.

According to Table 34 and assuming a per-capita allotment of 0.35 ha with a spring planting strategy, famines do not occur when wheat provides 55% or less of the caloric requirement. But when wheat provides 60% of the diet, famines can be expected 5% of the time. Famine frequency increases to 32% when wheat accounts for 65% of the diet, and the incidence of famines rises sharply thereafter. The outcome is somewhat different for a mixed strategy. As with spring planting, a mixed fall-spring planting allows only 55% of the diet to consist of wheat without leading to occasional famines. With wheat at 60% of the diet, famines can be expected 9% of the time, and with wheat providing 65% of the daily calories, famines can be expected 14% of the time.

The difference in the incidence of deficits can be traced to replanting problems discussed above. A series of years in which winter crops had to be replanted occurred during the first decade of the simulation. Whereas a spring-planting strategy allowed the village to survive marginally, a mixed fall-spring strategy resulted in a spiral in which stores were repeatedly exhausted to provide grain to reseed the winter crop. The spring planting strategy began to show small surpluses in sixth year, but the mixed fall-spring strategy did not show trade surpluses until the tenth year. Had the cropland been decreased instead of the seed borrowed from a neighboring village, the system would not have stabilized until several years later.

Once the mixed strategy stabilized, however, households maintained a larger wheat storage. The stores allowed the village to survive isolated years of poor harvests until a brief period, when replanting preceded, and was followed by, a poor harvest. Once again, the spring planting strategy was able to survive by drawing on household stores, while the mixed strategy entered a cycle in which deficits occurred because of the repeated deduction of seed grain from the household stores.

A mixed fall-spring strategy is risky because the village may enter a spiral of decreasing seed grain availability. The problem is visible for the 0.35 ha allotment, but it does not seem significant for the larger allotments. Prehistoric populations could be expected to adopt a mixed strategy only when large surpluses were needed and the socio-economic structure included a redistributive organization that ensured sufficient seed grain would be available in the event of several years of bad harvests. Such an organization did not appear in Central Europe until the Bronze Age.

A complicated relationship exists between the occurrence of village deficits, the maintenance of household stores, and the accumulation of trade surpluses. Table 33 shows the percentage of years in which trade surpluses existed for one or more houses, but it provides little information on the volume of surpluses. The purpose of this exercise is to determine whether the farmers could predictably provide the foragers with wheat. Therefore, dividing annual trade surpluses by the amount of wheat needed to fulfill the weekly food requirements of a hunter-gatherer band provides one mechanism for evaluating the relative merits of the trade surpluses. The caloric requirements of an arbitrary band of hunter-gatherers was established using the demographic profile suggested by Keene (1981:133; Weiss 1973: Table MT:15—50). The requirements are summarized in Table 36.

In order to evaluate the practical effect of each per-capita allotment and planting strategy, 50% of one week's caloric requirement was used to determine the number of weeks each set of combined strategies could provide a hunter-gatherer band with food. Table 37 and Figure 3 show the results of this evaluation.

Qualitative evaluation of trade surpluses suggests that effective surplus production is less than that projected in Table 33. Planting 0.35 ha per person, the village would be able to provide a band of hunter-gatherers with a 4-week supply of wheat only 61% of the time (if wheat fulfills only 55% of the cultivators' diet). However, because of extremely bountiful harvests, the settlement would be able to provide as much as a 16-week supply after 22% of the harvests under a spring strategy and after 38% under a mixed fall-spring planting strategy. When wheat provides 60% of the cultivators' diet, a one month surplus is accumulated after only 27% of the harvests of spring planted crops, and only 44% of the time with a mixed fall-spring strategy. The availability of trade surpluses drops sharply when wheat provides 65% of the cultivators' diet.

At 0.40 ha per person, a spring strategy provides more years with a 4-, 8-, or 12-week surplus than does the mixed strategy

Table 36

Reference Hunter-Gatherer Band

Age Cohort	Daily Caloric Require-ments	No. in Band	Net Caloric Require-ments
Infants	550	1	550
1–4	1360	4	5440
5–9	2010	4	8040
10–14 Girls	2420	2	4840
10–14 Boys	2750	2	5500
Women	2200	6	13200
Men	3000	6	18000
Total		25	55570

when the cultivators rely on wheat for 55% their annual diet, and a 4-week surplus when wheat supplies 60% of their diet. Both strategies produce the same number of years with an 8-week surplus, but a mixed strategy provides a greater number of years with surpluses of 12, 16, and 20+ weeks. When wheat provides 65% of the cultivators' diet, however, a mixed strategy provides more surpluses of 4, 8, 12, 16, and 20+ weeks.

If a 0.45 ha per-capita allotment is planted, a one-month supply is available 95% of the time when wheat provides 55% of the cultivators' daily caloric requirements; 88% of the time when wheat fulfills 60% of the diet; and 76% of the time when wheat provides 65%. The spring strategy more consistently provides a 4-, 8-, and 12-week supply than would a mixed planting strategy,

Table 37

Availability of Trade Surpluses for a Hunter-Gatherer Band

Ha Allotment	% of Wheat in Neolithic Diet	Planting Strategy	Number of Weeks (% of years)				
			4	8	12	16	20+
0.35	55	Spring	61	43	30	22	17
		Mixed	60	48	42	38	34
	60	Spring	27	16	9	4	3
		Mixed	44	37	31	25	18
	65	Spring	1	0	0	0	0
		Mixed	25	20	13	8	7
0.40	55	Spring	83	75	69	62	49
		Mixed	75	69	66	63	54
	60	Spring	69	63	51	33	26
		Mixed	67	63	57	47	42
	65	Spring	53	33	23	19	15
		Mixed	55	46	38	36	31
0.45	55	Spring	95	92	85	77	71
		Mixed	84	82	76	72	69
	60	Spring	88	80	73	68	64
		Mixed	78	73	69	67	63
	65	Spring	76	70	65	54	41
		Mixed	72	67	65	57	52

even though the mixed strategy provides more years with a 16- and a 20-week supply. The difference arising from spring and mixed fall-spring planting strategies becomes clearer when histograms of the surpluses are examined. As Figure 3 illustrates, spring planting strategies consistently result in smaller surpluses, whereas mixed fall-spring strategies have higher surpluses.

If meeting their own needs, participating in reciprocal trade with neighboring cultivators, and interacting with hunter-gatherers were goals of the cultivation system, then it would be more important to avoid deficits and consistently produce a 4-, 8-, or 12-week surplus, than to occasionally have years with extreme surpluses. With wheat providing 55–65% of the diet, and planting 0.40–0.45 ha per person, a spring planting strategy provided at least as many 4-, 8-, or 12-week surpluses as did the mixed fall-spring strategy. A spring planting strategy was at least as satisfactory as a mixed fall-spring strategy and it was less risky. The 0.45 allocation produced a superabundance of wheat in the majority of years. An allotment of either 0.35 or 0.40 ha per person appears to produce sufficiently large harvests of wheat to provide 55–65% of the diet. Therefore, only the 0.35 and the 0.40 ha allotments will be considered when stockbreeding is examined.

Stockbreeding

Thus far the discussion has concentrated on wheat production; however, cereal crops can serve several purposes. Straw can be fed to livestock and it can be used for roofing material, as well as for baskets, hats, and other utensils. The simulation estimated the number of days each harvest would provide winter fodder for the ruminants. The needs of a 30-head herd of cattle were deducted first, and if straw remained, then the requirements of both the ovicaprids were deducted for goat herds from 15 to 50 head and sheep flocks up to 50 head. The calculations were repeated for a 40-head cattle herd, and the various ovicaprid herd sizes; then for a 50-head cattle herd and, once again, the various ovicaprid herds. Tables 38–39 show the frequency with which harvests would provide sufficient winter fodder for the ovicaprid herds, having first deducted the requisite amounts for the cattle herds.

Figure 3a. Histograms of Trade Surpluses: A. 0.35 ha allotment

Figure 3b. Histograms of Trade Surpluses: B. 0.40 ha allotment

Table 38

Harvests Providing Sufficient Winter Fodder: 0.35 ha/per capita
(% of years)

Cattle Herd Size	Size of Combined Ovicaprid Herds							
	30	40	50	60	70	80	90	100
30	96	94	87	84	68	57	38	30
40	20	12	7	3	1	–	–	–
50	–	–	–	–	–	–	–	–

Table 38 shows the results for an allocation of 0.35 ha per-capita arranged with reference to cattle herd sizes. The table indicates that straw would optimally provide winter fodder for a 30-head cattle herd plus 60, 70, or 80 goats and sheep. If a 70-head mixed ovicaprid herd is used as an optimal figure, then extra hay would be needed for ovicaprids after 32% of the harvests.

Table 39 presents the results of a per-capita allocation of 0.40 ha. While harvests could occasionally provide sufficient fodder for 50-head cattle herds, they would more consistently provide winter fodder for a 40-head cattle herd plus a mixed ovicaprid herd. If a combined ovicaprid herd of 40 (20 goats and 20 sheep) is taken as the optimal figure, then additional hay would be required after 33% of the harvests. But, 34% of the harvests would provide surplus straw, which could be used as roofing materials or for manufacturing tools and utensils.

A per-capita allotment of 0.40 ha provides sufficient, but not excessive, harvests of wheat and adequate straw yields. Estimates of optimal resource mixes will therefore be based on an allotment of 0.40 ha per person, with 40 cattle, 20 goats, and 20 sheep. Adopting these figures allows an optimal model of Early Neolithic cultivation and stockbreeding to be constructed.

Table 39

Harvests Providing Sufficient Winter Fodder: 0.40 ha/per capita
(% of years)

Cattle Herd Size	Size of Combined Ovicaprid Herds							
	30	40	50	60	70	80	90	100
30	99	99	97	96	96	94	92	89
40	69	67	43	35	31	23	9	6
50	5	2	1	1	–	–	–	–

Optimal Resource Mixes

As concluded from the simulation above, a per-capita allotment of 0.40 ha will be planted per person (excluding infants). Assuming a 40-head cattle herd, 16 cows would have been divided between the six households and results in an average of 2.6 milk cows per household[2]. The additional yearlings, heifers, and steers would be evenly distributed throughout the village (Table 40). Because of their larger fields, the largest households would produce more straw; therefore, they would probably be better able to maintain the bulls. The two largest households have each been assigned one bull.

A 20-head herd of goats would provide 9 milking does, while a 20-head flock of sheep would provide 9 milking ewes. This divides neatly into the six households, but it is doubtful all households

[2]Interestingly, the figure is very close to average of 2.5 that Netting (1981) found to be consistent for a 125 year period in a modern-day peasant Alpine village.

Table 40

Livestock Distribution Throughout the Reference Village
(winter herd/flock composition)

Household	Cattle			Goats			Sheep			Pigs	
	Cows	Steers, Heifers & Yearlings	Bulls	Yearlings	Does	Buck	Yearlings	Ewes	Rams	Sows	Boars
A.	2	3		1	2	1	1	2			
B.	3	4		1	3		1	3	1		
C.	3	4		1	2		1	3	1		
D.	2	3		1	2	1		2			
E.	3	4	1	1	2			2		1	1
F.	3	4	1		2		1	2		1	

had three milking ovicaprids every year. More likely, some households would have two, some three, or others four.

Pigs reproduce quickly, and they provide an excellent source of meat and fat. Unlike cattle and ovicaprids, they provide no renewable resources; however, they do provide a valuable service. Pigs are omnivorous and consume carrion, rotting vegetable matter, and human excrement. Their importance as scavengers should not be minimized: pigs can play an important role in reducing village filth. Large herds could be kept, but pigs would have competed with the humans for food. In this model, pigs are treated as a highly elastic resource, because of their high reproductive rate. In poor years the herd could be heavily culled with little effect on future productivity. In normal years, pigs need not be provided with food other than table scraps, household garbage, winnowing and threshing debris, what they can scavenge from the village, and acorns that may have been collected. After the harvest, pigs could be allowed to root in the fields, and following bumper crops, they could have been fattened on surplus cereals throughout the winter.

Meat and Milk Production

The optimal flock and herd sizes allow annual estimates of meat and milk yields to be established. Beef, goat meat, mutton, and pork vary in nutritional quality. Gram-per-gram, protein provides about half the calories that fat provides. Thus the caloric content of meat depends primarily on the proportion of fat. Pigs have the largest proportion of fat, whereas goats have the least. Beef and mutton fall between the two extremes. The nutritional value of milk also depends to a large extent on its fat content. Sheep milk has the highest fat content, and cow milk the least. Moreover, the fat globules of both goat and sheep milk are much smaller than those found in cow milk, and both goat and sheep milk are much easier to digest.

As shown in Tables 41–42, estimates of the nutritional value of the different meats and milks vary from author to author. The variation arises from differences in breeds selected for analysis, the condition of the animals, and test procedures. There is probably no single "correct" figure for the nutritional value of the meats and milks, but this may be a moot point, for the breeds of prehistoric livestock are unknown. The ultimate selection of

Table 41

Nutritional Value of Beef, Mutton, Goat Meat and Pork (per 100 gram portion)

	Water (%)	Energy (cal)	Protein (g)	Fat (g)
Beef				
Watt & Merrill	62.5	242.0	18.6	18.0
Goat Meat				
Pellett & Shadarevian	71.5	157.0	18.4	9.2
FAO		123.0	14.0	7.0
Ensminger *et al.*	71.0	165.0	18.7	9.4
Mutton				
Pellett & Shadarevian	61.0	257.0†	17.0	21.0
FAO		241.0	11.9	21.1
Watt & Merrill	62.5	247.0	16.8	19.4
Pork				
Watt & Merrill	56.4	245.0	27.6	14.1
Ensminger *et al.*	56.4	245.0	27.6	14.1

Sources: Ensminger *et al.* 1983; FAO 1953; Pellett and Shadarevian 1970; Watt and Merrill 1975.

†The figure 257 is given in Pellett and Shadarevian 1970: Section I. According to Redding (1981) appears to be a typographical error. When the grams of protein and fat are multiplied respectively by 4 and 9 grams, the caloric value of 257 is obtained.

Table 42

Nutritional Value of Sheep, Goat, and Cow Milk
(per 100 gram portion)

	Water (%)	Energy (cal)	Protein (g)	Fat (g)
Cow Milk				
Pellett & Shadarevian		64.0	3.5	3.0
FAO		65.0	3.5	3.5
Ensminger *et al.*	87.2	66.0	3.3	3.7
Watt & Merrill		65.0	3.5	3.5
Goat Milk				
Pellett & Shadarevian	87.0	70.0	3.3	4.0
FAO		73.0	3.8	4.5
Ensminger *et al.*	87.5	67.0	3.3	4.0
Watt & Merrill	87.5	67.0	3.2	4.0
Sheep Milk				
FAO		99.0	5.8	6.5
Ensminger *et al.*	80.7	108.0	6.0	7.0

Sources: Ensminger *et al.* 1983; FAO 1953; Pellett and Shadarevian 1970; Watt and Merrill 1975

nutritional values to use rested on the need to remain as consistent as possible in constructing the model. No source provided estimates for all meats and milks, and it was not possible to use a single reference for all estimates. Redding (1981) suggests that Pellett and Shadarevian (1970) provide the most reliable estimates for both goat and sheep meat, so their estimates have been used for the ovicaprids.

One problem in using Western estimates for beef and pork is that these animals are now fattened on high-quality cereals or legumes, and a high level of fat in the meat results. Watt and

Merrill (1975) give estimates for utility grade beef, a grade with a lower fat content than higher grade cuts. Utility-grade estimates were used for beef in the hope that the values would approximate more closely the fat content of unimproved breeds. The meat offtakes for a 40-head herd of cattle, a 20-head herd of goats, and a 20-head flock of sheep are given in Tables 23, 31, and 32 respectively. As discussed earlier, each of two pig litters would result in 40 kg meat for a total of 80 kg pork. Multiplying these amounts by the caloric value of beef, mutton, goat meat, and pork respectively, livestock would provide an annual 4,289,245 calories. This amounts to 15.3% of the settlement's annual caloric requirements.

Ensminger *et al.* (1983) provide nutritional information for cow's, goat's, and sheep's milk. The values Ensminger *et al.* provide for sheep's milk is somewhat higher than that suggested by the FAO (1953), but their values for goat's and for cow's milk agree closely with those provided by Watt and Merrill (1975). Their cow's milk value is slightly higher than that suggested by FAO (1953) and Pellett and Shadarevian (1970). Because they provide figures for sheep, goats, and cattle, and because the values are roughly equivalent with estimates provide by other nutritional studies (with the exception of sheep milk), the figures provided by Ensiminger *et al.* are used in calculating the caloric content of the milk produced by the reference herds. If the expected annual milk yields from Tables 23, 30, and 31 are multiplied by the figures suggested by Ensminger *et al.*, milk could be expected to provide 2,186,943 calories annually. This fulfills 7.7% of the annual diet.

If it is assumed that each person consumes 20 grams of legumes daily, as suggested for cereal cultivators in India (Sen Gupta 1980), the village would require 240 kg of peas and lentils annually. Divided equally among peas and lentils, this amounts to 120 kg of each. Using the figures provided in Table 4, 120 kg of lentils provides 408,000 calories and 120 kg of peas provides 417,600. Combined, the legumes would fulfill 2.9% of the annual requirements.

Table 43 sums estimates of cereals, meat, milk, and legumes. Wheat is assumed to provide 60% of the diet, while all other dietary contributions are as discussed above. From the table, it is evident that cereal cultivation and stockbreeding could provide 85.9% of the annual caloric needs. A further 14.1% of the annual diet must be derived from wild foodstuffs, or the significance of cereal or legumes in the cultivator's diets must be increased.

Table 43

Fulfilling Reference Village Dietary Needs

	Calories	Percent of Annual Requirements
Annual Requirements	28,090,800	
Wheat	16,841,877	60.0
Meat	4,289,245	15.3
Milk	2,186,943	7.7
Legumes	825,600	2.9
Total		85.9

If the role of cereals in the diet is increased by 5%, the village should not experience any famines; however, the number of years the village could provide hunter-gatherers with a minimum of a 4-week surplus would drop by 23.18%, from 69 to 53 years out of 100. Perhaps more significant, the number of years in which they could provide an 8-week supply would decrease by 47.62%, from 63 to 33 years. Increasing the per-capita allotment for wheat does not seem a logical alternative, for this would result in a superabundance of wheat in the majority of years, even if substantial amounts were given to the foragers. The systematic inclusion of wild foodstuffs in the subsistence strategy of the farmers seems the best alternative. Examining the seasonal activities of the production year in conjunction with the dietary requirements of the consumption year shows the degree to which this may have been possible. The next section discusses the seasonal work schedule; a detailed model of wild resource exploitation will be developed in Chapter 7.

Chapter 6

Subsistence Activities

Planting and harvesting are the two most time consuming activities in the cultivation year, and each activity must be completed within a relatively short period. Estimates of these activities provide the starting point for determining the seasonal scheduling of cultivation activities.

Planting

Historical records from southwest Germany were consulted to determine the scheduling of spring planting activities (Statistisch-Topographisches Bureau 1850–1905). Figure 4 provides a schematic diagram of the scheduling of planting activities. All crops were planted as soon as the soil could be prepared; however, planting dates varied up to three weeks depending on spring weather. Wheat was always given priority and planted first, while peas, poppies, flax, and lentils followed in short order. Planting all four crops required unbroken labor for several weeks. The results of cropping experiments provide data for estimating the amount of time that prehistoric cultivators would have needed to plant their crops.

Steensberg (1979) notes that in his experiments of growing wheat, each square meter required 30 minutes for hoeing and planting. Extrapolating this figure to one hectare, 500 hours labor would be required for preparing the land and sowing the wheat. Dividing this figure into eight-hour work days, one hectare would require 62.5 days of labor. This seems excessive. Steensberg planted only a limited area, so there was little time pressure. His workers may have been unduly thorough because of the limited area, and the results may be skewed. If 15 minutes' preparation and planting time is allowed for each square meter, then one hectare would require 31.25 labor days. The labor requirements for each household can be estimated by multiplying each household's cropland requirements by this figure and then dividing by the number of older children and adults in the household. These figures are summarized in Table 44.

Table 44 suggests problems may arise in planting the wheat crop. The ground must not be sodden when wheat is planted, for wet soil lowers the germination rate; moreover, there is a strong likelihood that the seedlings will become waterlogged. Thus, cul-

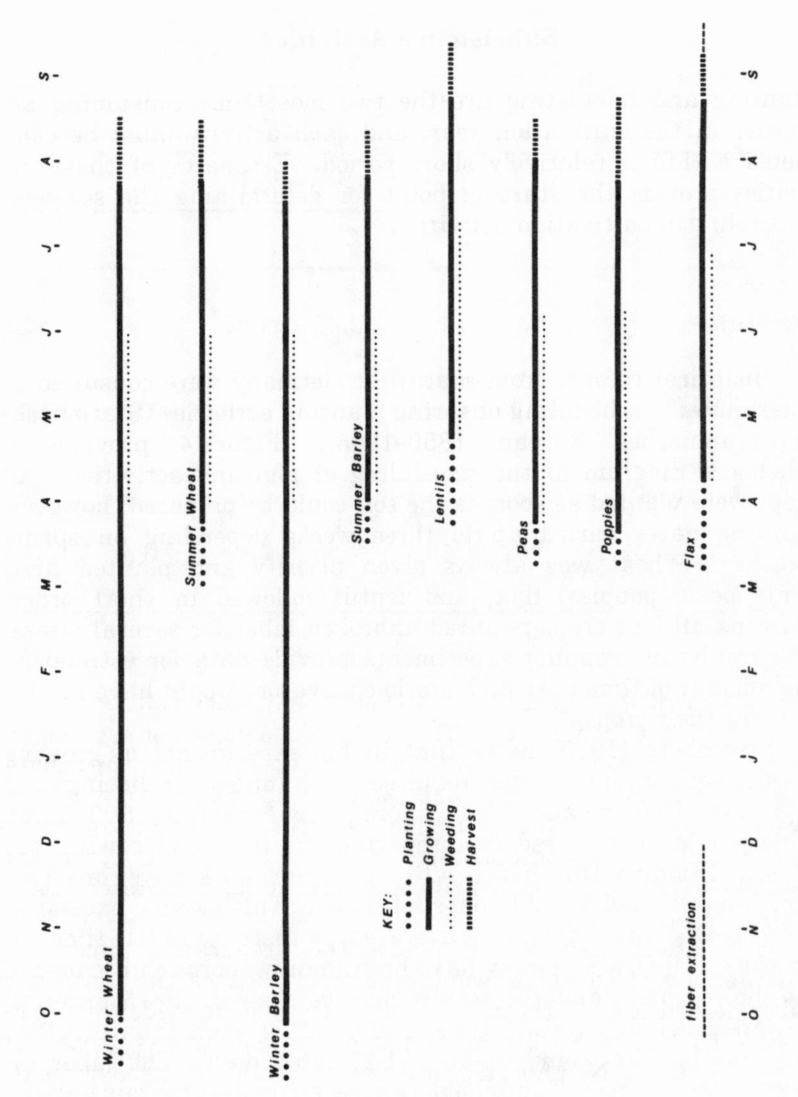

Figure 4. Cultivation and Harvest Schedule

Table 44

Labor Requirements: Wheat Planting

	Household					
	A	B	C	D	E	F
Hectares	1.60	2.00	2.00	1.60	3.20	2.80
Work Days required	50.00	62.50	62.50	50.00	100.00	87.50
Workers per Household	2	3	3	2	6	5
Net Work Days	25.0	20.8	20.8	25.0	16.6	17.5

tivators must wait until the soil is sufficiently dry before they can begin hoeing and planting. The worst consequence of such delays is that they shorten the tillering period, and this contributes directly to a lower harvest. Late, wet springs can mean a delay of two or three weeks. If a household requires a further three weeks to complete hoeing and planting, yields can be reduced significantly and the peas, poppies, lentils, and flax would be planted late as well.

Two alternatives would reduce the preparation time. The introduction of the plow would effectively reduce the amount of labor needed to turn the soil; however, there is no evidence of plow before the Middle Neolithic. The second alternative is to increase the labor force. Neighboring cultivators would be faced with the same time constraints, and they would be affected by the same weather conditions. Thus it is unlikely that neighboring settlements would provide labor. Hunter-gatherers, however, were a potential labor pool. Ethnographic examples of hunter-gatherers working during the planting season are known from Southeast Asia (Gregg 1981; Pookajorn 1982), and this may have occurred in prehistoric Europe.

Table 45

Net Planting Time with Four Hunter-Gatherers

	Household					
	A	B	C	D	E	F
Work Days Required	50.00	62.50	62.50	50.00	100.00	87.50
Workers per Household	2	3	3	2	6	5
Additional Workers	1	1	1	1	0	0
Net Labor Force	3	4	4	3	6	5
Net Work Days	16.6	15.6	15.6	16.6	16.6	17.5

Additional labor provided by the hunter-gatherers could significantly reduce the amount of time needed to cultivate the land and sow the wheat. The reference hunter-gatherer band could potentially provide 16 additional laborers (six women, six men, and four older children). If only four hunter-gatherers are added to the cultivators' work force, then all households could plant their crop within slightly more than two weeks (Table 45). Furthermore, the hunter-gatherers could continue to pursue their own subsistence activities and could provide the remaining 50% of the requisite food. If all of the adult hunter-gatherers worked, then the wheat could be planted in less than two weeks (Table 46); however, the farmers would have to provide all food for the foragers. The reduction in planting may not be important for most years, but it would be absolutely critical for years in which the spring was late.

Peas and lentils require substantially less labor than does wheat. If each person consumed 20 grams of legumes per day,

Table 46

Net Planting Time with Twelve Hunter-Gatherers

	Household					
	A	B	C	D	E	F
Work Days Required	49.76	62.20	62.20	49.76	99.52	87.08
Workers per Household	2	3	3	2	6	5
Additional Workers	2	2	2	2	2	2
Net Labor Force	4	5	5	4	8	7
Net Work Days	12.4	12.4	12.4	12.4	12.4	12.4

each individual would require 7.3 kg of lentils and of peas. Assuming an average yield of 1473.40 kg/ha for peas and 1152.00 kg/ha for lentils (see Table 5), each person needed 5.0 m^2 and 6.3 m^2 for peas and lentils respectively. Multiplying these figures by the household size gives the area of the plot needed to fulfill each household's legume requirements. Once again assuming 15 minutes per square meter for soil cultivation and planting and an 8-hour work day, the amount of labor can be estimated (Table 47). Dividing the work requirement by the number of adult women in each household shows that less than two days would be needed to plant the pulses.

At present the significance of linseed and poppy seed in the Neolithic diet, and the requirements for flax in producing cloth, cannot be determined. Therefore, additional estimates of labor requirements needed to plant flax and poppy will not be made.

Table 47

Labor Requirements: Legume Planting

	Household					
	A	B	C	D	E	F
Plot Size (m^2)	45.6	57.0	57.0	45.6	91.2	79.8
Work Days Required	1.42	1.78	1.78	1.42	2.85	2.49
Females per Household	1	1	1	1	3	3
New Work Days	1.42	1.78	1.78	1.42	0.95	0.83

Harvest

In general, wheat ripens in mid- to late August, followed by flax in late August to early September, and both peas and lentils in early September. The poppies can ripen slightly before, or at the same time as, wheat. The timing of the harvest is critical, for severe storms may result in delays. Since mouse predation increases in severity as the crop ripens, delays in harvesting can result in increased losses due to mice. It is imperative, therefore, that wheat be harvested as soon as it is ready. Steensberg (1979) provides figures on the amount of time needed to harvest wheat using flint sickles. His figures suggest that 32.1 hours are needed to harvest one hectare of wheat, but this does not include transporting the wheat to the threshing place, nor does it include threshing and winnowing. Table 48 estimates the amount of time each household would require for harvesting only.

There is at present no evidence for the use of wheeled vehicles in the Early Neolithic; horses had not been domesticated. Cattle do not appear to have been used as beasts of burden, but they could have been used to transport wheat to the threshing place. Dogs may have also been used to pull travois. There is no evidence for either, so it is assumed that human labor was used. This produces a conservative estimate of the amount of the time needed to transport the wheat. Men could have carried an average of 30 kg, and women 20 kg per trip. The harvests in the simulation weighed up to 1000 kg for the small households to 1900 kg for the large households. If each round trip required 30 minutes, then the smaller households would require up to 20 round trips or a further 10 hours of labor. The larger households would require up to 38 round trips for a further 19 hours of labor. The larger households have twice the number of adults, so they could transport their wheat within the same number of days. This would add slightly more than one day's labor to the harvest estimates given in Table 48.

Table 48

Labor Requirements: Wheat Harvest

	Household					
	A	B	C	D	E	F
Hectares	1.60	2.00	2.00	1.60	3.20	2.80
Work Days	6.42	8.03	8.03	6.42	12.84	11.24
Adults in Household	2	3	3	2	4	4
Net Work Days	3.21	2.68	2.68	3.21	3.21	2.81

A further two to three days should be added for threshing and primary crop cleaning for each household. Even adding these days to the estimated harvest and transport figures suggests that each household would have had sufficient labor for harvesting its wheat. Additional labor might have been welcome, but, it probably would not have been critical.

Two final considerations should be made in estimating baseline subsistence activities. These are the amount of wood needed for fuel and for construction purposes.

Fuel and Shelter

Fuel would have been needed daily for cooking, and for heat throughout the winter. Estimating the amount of fuel needed by an Early Neolithic settlement is complicated by several factors. First, reconstructions of Early Neolithic houses (Mausch and Ziessow 1985; Meyer-Christian 1976; Startin 1978) acknowledge that ovens may have been used, but they do not include fireplaces with a chimney. Fires may have been on open hearths, with smoke rising into, and escaping through, the thatched roof. Smoke circulating through the thatch would have greatly reduced the amount of vermin in the roof, and this would have extended its use-life. However an open hearth is ineffective in heating, and in the winter it would have required large quantities of wood.

When wood burns, the fire goes through three stages. First, moisture is driven out of the wood. Second, the wood breaks down chemically, and volatile matter is vaporized. Third, the charcoal remaining after step two is burned. Fresh cut wood has a 30% moisture content, so green wood should be allowed to dry for several months before it is burned (Harris 1980); otherwise, the wood burns at a lower temperature, and more wood is needed to obtain the same amount of heat. Trees contain less water in the winter, the lack of leaves allows the woodsman to select trees more carefully, and a cover of snow helps in dragging trees from the forest. Therefore, late fall and winter are optimal seasons for cutting and hauling firewood. The burning value of wood varies from species to species. If wood with an average or poor burning quality is used, then larger amounts are needed to obtain the same heat. Using wood with a high burning value generally results in the most economic use of fuel. As shown in Table 49, apple, beech, and oak provide the best fuel; they have little smoke and few sparks. But they are difficult to split, and they split easiest when green. Optimally, splitting should be completed shortly after the tree is felled.

Table 49

Fuel Values

Species	Burning Value	Starting Ease	Sparks	Smoke	Splitting	Comments
Apple	Excellent	Poor	Few	No	Tough	Fragrant
Beech	Excellent	Poor	Few	No	Tough	Seasons well
Oak	Excellent	Poor	Few	No	Tough	Split when green
Ash	Average	Fair	Few	No	Fair	Burns well when green
Birch	Average	Good	Some	No	Fair	Must split to avoid rotting
Elm	Average	Fair	Very few	Some	Very tough	Fire must be fed
Maple	Average	Poor	Few	No	Tough	
Pine	Poor	Excellent	Many	Some	Easy	Quick, hot fire
Willow	Poor	Fair	Few	n.d.	Fair	

Sources: Harris 1980; Shelton and Shapiro 1976

A very rough estimate of the needed amount of fuel can be obtained through comparison to contemporary fuel requirements for wood burning stoves. In New England a family heating their home with wood requires 20 cords to get through the winter. Modern wood stoves are highly efficient, particularly in comparison to open-hearth fires. Thus, wood requirements in the Early Neolithic may have been significantly higher. Counterbalancing this, Early Neolithic winters would have been milder than a typical New England winter. Moreover, animals were undoubtedly stabled in close proximity to human living quarters, and the humans would have benefitted from the animals' warmth. Finally, fire would have been used primarily to heat the main living room. Lacking detailed information on the energy efficiency of Early Neolithic houses, 20 cords will be used as a baseline in determining fuel requirements. A minimum of 2 cords would be needed for cooking fires throughout the rest of the year. A well managed woodlot can continuously provide 2.5 cords/ha (Harris 1980). Thus each house would require an 8.8 ha woodlot, and the settlement would require a total of 52.8 ha. If wood with average or poor burning qualities were used, the annual wood requirements would be greater.

Construction Materials

If a house had a 30-year use life and the village had six households, then a new house would be constructed on the average of once every five years. Startin (1978:154) suggests that constructing a long house would require all the trees from a 0.40 ha area. Assuming the cleared area was allowed to return to forest, regeneration would be complete within a 60-year period. Over six decades, the model village would require a forested area of 4.8 ha. In all, a six-household village would require a 57.6 ha woodlot to fulfill their requirements for construction timbers and fuel.

Summary and Conclusions

Results of the simulation suggest a six-household village would need to plant 13.2 ha of wheat and that the village could maintain a 40-head herd of cattle, plus a combined herd of 40 ovicaprids. A general estimate of the village's impact on the landscape can now be developed (Fig. 5). The hypothetical village

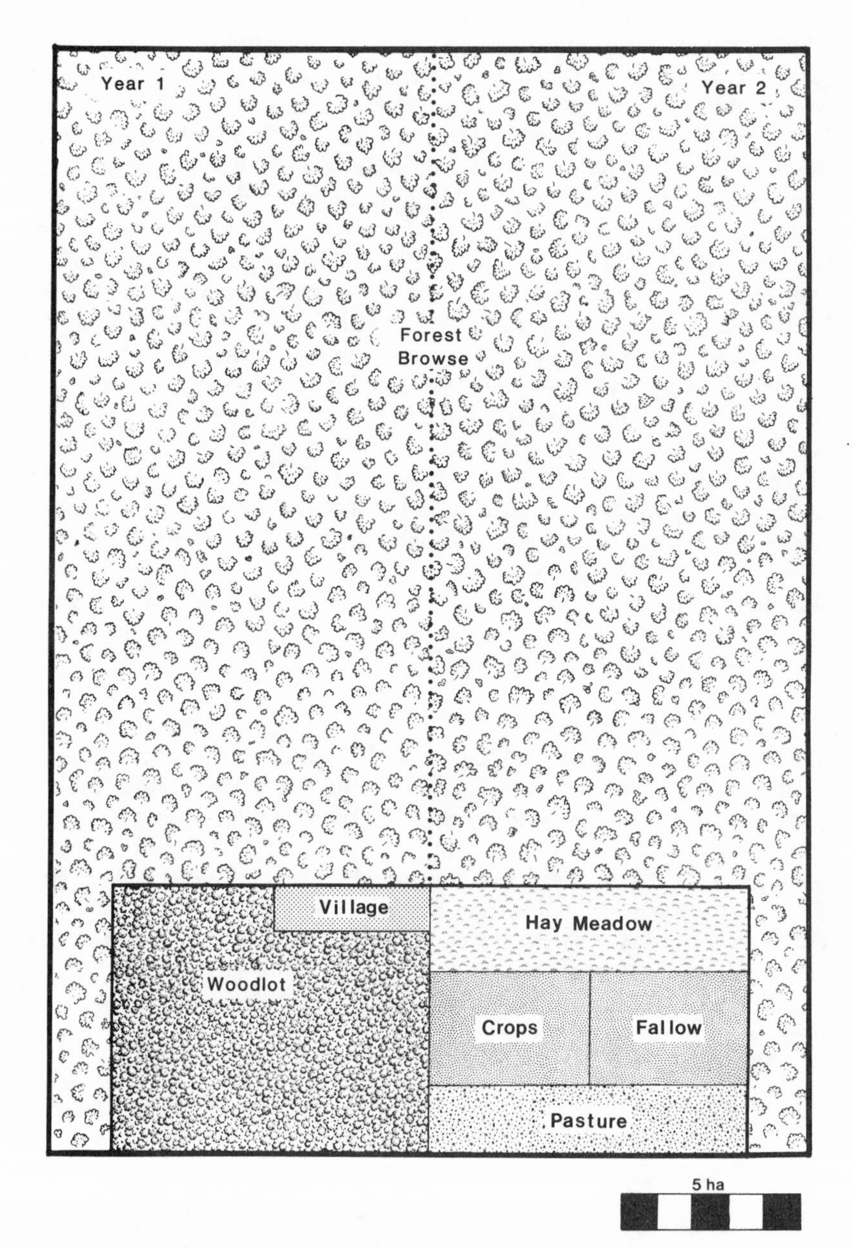

Figure 5. Schematic Diagram of Neolithic Land Use

would require approximately 4.5 ha for the houses, outbuildings, and garden plots (for legumes and other crops) and the woodlot would cover 52.8 ha, with a further 4.8 ha needed for construction materials. The livestock would require 18.18 ha of pastureland, 19.66 ha of natural meadows, and 2.56 km^2 for forest browse. Because of the degradation that would occur if the forest were overgrazed, the forest browse figure minimally should be doubled to allow for a two year rotation in forest use. Although soil fertility would not be seriously depleted with grain yields of less than 1000 kg/ha, archaeological weed assemblages do not suggest that communities of weeds associated with cereals cultivated in continuously used fields were present in the Early Neolithic (Willerding 1981). This suggests that a fallow of minimally one year occurred.

The reference village would thus require a minimum of 6.07 km^2 to obtain their domestic resources. Figure 5 provides a schematic diagram of the reference village land requirements. The actual requirements of a village would, of course, vary with the specific terrain in which the village is situated. From Figure 5 it is apparent the livestock requirements, not the cultivation activities, account for the greatest impact on the deciduous forest.

The farmers would be able to produce wheat surpluses that could have been traded to the foragers, who, in turn, had two potential commodities which the farmers may have required. The first was labor during the late winter and early spring. The second may have been a supply of wild foodstuffs during the early summer prior to the cereal harvest. The degree to which the farmers may have competed with the foragers and whether or not the hunter-gatherers would have participated in such an exchange is considered in the next chapter.

7

Wild Resource Exploitation: Competition, Cooperation, and Interaction

The past three chapters evaluated constraints affecting farming and developed an optimal model of Neolithic farming and stockbreeding. Focus shifts in this chapter to identify potential competition between the foragers and the farmers. The amount of wild resources needed in the Neolithic diet would have been established by the optimal cultivation strategy and the size of livestock herds; similarly, the quantity of wild resources in the Mesolithic diet would have been determined by the quantity of domestic resources they obtained from the farmers. Interaction would contribute varying levels of domestic resources to the foragers' diet, thereby reducing the requisite amount of wild resources and leading to further modifications in resource use.

To fully explore potential areas of competition, the effects that varying levels of domestic foodstuffs would have had on the exploitation patterns of each population must be identified. Evaluating the significance of wild resources in a domestic economy and domestic resources in a foraging economy can be incorporated into a single a model that investigates domestic and wild resource exploitation.

The Resources

A predictive model of Mesolithic subsistence strategies has already been developed for the study area (Jochim 1976), and it provides a simple but elegant means for examining the structure of wild resource exploitation. The model is based on the monthly use of each resource as part of a total annual diet. It is relatively simple to add additional resources. Expanded and revised to include domestic resources, the model allows a month-by-month comparison of Neolithic and Mesolithic resource use. By examining changes resulting from different proportions of domestic and wild resources in their respective diets, it is possible to identify the effects that different resource configurations might have on competitive or cooperative relationships. In order to determine the monthly requirements of each population, two calculations must be made for each resource: (1) its proportional contribution to the *annual* diet; and (2) its *monthly* use. Before these calculations can be made, the relative importance of all wild and domestic resources must be established. The availability of domestic resources depends on the subsistence strategies of the farmers. The relative importance of *each* wild resource, however, is affected by the nutritional value and the availability of *all other* wild resources.

Before the model is explained, the wild resources are briefly introduced. The purpose in developing the model is to identify salient features affecting wild resource exploitation during the Mesolithic/Neolithic transition, rather than to reconstruct the hunting strategies practiced at a specific site. In order to simplify the problem, only the animals most commonly appearing in Mesolithic and Neolithic contexts are incorporated in the present model, and rarely occurring species are excluded. Since remains of aurochs (*Bos primigenius*), forest bisons (*Bison bonasus*), brown bears (*Ursus arctos*), and wolves (*Canis lupus*) are uncommon (Boessneck *et al.* 1963; Bogucki 1982), it is assumed they were not hunted on a regular basis, and they will not be included in the present model. Eight wild resources commonly occurring at Mesolithic and Neolithic sites are included. Four of these are treated as separate resources: red deer (*Cervus elaphus*), roe deer (*Capreolus capreolus*), boar (*Sus scrofa scrofa*), and beaver (*Castor fiber*). Fish, birds, small game, and wild plant foods are grouped into four collective classes. The seasonal availability and the density of each resource are critical variables; therefore, the ecology of the wild resources is presented below.

Wild Game

Red Deer

European red deer are comparable to North American elk, and many ecologists no longer treat the two as separate species (Bryant and Maser 1982; *cf.* Clutton-Brock *et al.* 1982). Red deer tolerate a wide variety of environments, and a number of subspecies have been recognized based on their adaptation to specific environmental conditions. They are gregarious animals and spend most of their time in segregated bands of stags and of hinds along with their young. Group membership fluctuates considerably. In one study, the median party size for hinds (excluding their young) in summer is reported as 6.88, and in winter, 5.24; for stags, the median party size in summer is 6.92 and in winter 4.0 (Clutton-Brock *et al.* 1982:181). In rutting season stags frequently have harems of 4 to 5 females, but harems of up to 52 females have been reported (Clutton-Brock *et al.* 1982:180). Like other ruminants, red deer depend on a bulky diet consisting of grasses, herbs, and shrubs. Red deer migrate in the spring, when herds move out of their lowland winter quarters to their upland summer range, and then again in the fall as they return to protected areas for the winter.

In North America cattle and elk have a competitive relationship (Nelson 1982). Both prefer grasses in their diet, and heavy grazing by one species can adversely affect the other. Competition appears to be most serious on the winter range, and cattle have the dominant role. Ecologists are now beginning to realize that the two species are socially incompatible, and this incompatibility appears to be a significant factor in the competition between them. Because of both social incompatibility and direct competition for graze, rates of elk use decrease as cattle usage increases (Nelson 1982). Sheep also compete with elk, although not as intensely as cattle. Competition with sheep occurs primarily in high summer ranges.

Red deer densities vary depending on the amount of forest cover, the quality of forage, and the amount of competition from other ruminants. Jochim (1976:101) suggests a density of 4 per square kilometer, and this figure has been adopted as a standard density. Competition with cattle and sheep can result in lower red deer densities, which would influence the availability of red deer as a resource. If the livestock needed several square kilometers for browsing, as suggested in Chapter 6, a lowered red deer density should be expected near a village. To account for the potential effects of stockbreeding, a lowered red deer density of two individuals per square kilometer was also investigated. This is a much lower density than Jochim favors, but it is higher than the prehistoric elk density suggested for comparable deciduous forests of mid-western North America (see Keene 1981:105).

Roe Deer

Roe deer are not as gregarious as red deer. Instead, two or three individuals browse together. Shrubs provide the bulk of the roe deer diet. These deer require territories with a well-developed shrub layer, and they favor ecotones formed along forest margins. Jochim (1976:103) suggests a density of 12 individuals per square kilometer. It was argued above that lowered red deer densities should be expected as a result of Neolithic stockbreeding activities, but it might be contended that the same activities raised roe deer densities. Clearing forested land would have resulted in a disproportionately large amount of forest margin in areas around a village, which would have favored an increased density of roe deer. Concomitantly, livestock browsing and grazing would have reduced the amount of available forage. It has been assumed that the two would have counter-balanced one another, and the roe deer densities have been kept at 12.

Boar

Wild boar prefer closed forests. Unlike their modern domestic counterparts, they are fast, sleek animals well suited to moving through dense undergrowth. Wild boar are omnivorous and consume a variety of roots, herbs, grasses, and decaying organic matter. Nut crops are critical fall resources that allow boar to add several kilograms of body weight prior to the onset of winter. Jochim conservatively suggest a density of 12 per square kilometer.

Beaver

Beaver live either in small lodges built in ponds, or in burrows dug into stream or banks. They are social animals that form stable communities, and the average family group has five members. Beaver feed on the bark and twigs of willow, cottonwood, birch, maple, and ash as well as on the roots and rhizomes of sedges and rushes. Their body weight and composition is stable throughout the year (Patrick and Webb 1960), and their pelts are thick and luxurious in winter. Jochim suggests a density estimate of 0.64 per square kilometer.

Fish, Birds, and Small Game

The study area includes a glacial landscape with an abundance of small lakes. The Danube River (which is the size of a very small stream) cuts across its northern portion. The region is not on a flyway, so there is no super abundance of fowl during either the fall or spring migration periods; nonetheless ducks, geese, mudhens, and other waterfowl are available throughout the summer. Grouse are present throughout the year. Wild fowl would have been available, but not in dense quantities. Perch, pike, trout, chub, and catfish are common in the lakes and streams; moreover, several species of cyprinids spawn in large schools. Fish would have been an important resource, although there are no anadromous fish runs in the area. A variety of small fur-bearing animals would have been readily available, including squirrels (*Sciurus vulgaris*), rabbits (*Lepus europaeus*), otters (*Lutra* sp.), martins (*Martes* sp.), wild cats (*Felis silvestris*), hedgehogs (*Erinaceus europaeus*), and badgers (*Meles* sp.). The animals would have been important both as a food resource and for their furs.

Jochim suggests a collective density of 98 kilograms per square kilometer for fish and 103 individuals per square kilometer for small game. Rather than estimate a density for birds, he suggests that birds fulfilled 2% of the Mesolithic diet. Jochim's density figures were adopted for the foragers; however, for the farmers, birds were lumped with small game to form a single class of resources.

Wild Plant Foods

The exploitation of wild plant foods is one of the least understood aspects of the Mesolithic and Early Neolithic subsistence strategies in Central Europe. Ethnographic and archaeological studies in North America demonstrate the importance of wild plant foods to horticulturalists (Cowan 1985; Ford 1978; Yarnell 1964). Excavations at waterlogged Mesolithic sites in Northern Europe and Middle Neolithic sites in Central Europe have produced a plethora of uncharred nuts and seeds from grasses, berries, and fruits, and it is generally assumed that wild plant foods were a regular component of their subsistence strategies. In all likelihood, wild plant foods were important for both Mesolithic and Early Neolithic populations. In archaeological contexts, the absence of evidence cannot be taken as evidence of absence; therefore, the lack of wild plant remains in Mesolithic and Early Neolithic contexts should not be interpreted as proof that plants were excluded from the set of exploited wild resources. Nuts, berries, seeds, greens, and tubers were undoubtedly exploited. The following discussion reviews pertinent characteristics of each class and considers their potential contribution to the human diet.

Nuts

Nuts are one of the few wild plant resources for which there is evidence of use. Only hazel shrubs and oak trees provided nuts until the widespread dominance of beech trees—which marks the start of the Middle Neolithic—and chestnut trees during the Roman period. Hazel nut shells occur in charred floral assemblages from Mesolithic sites, but acorn shells do not. Because of this, archaeologists assume that acorns were not exploited. Acorns are not included in this discussion.

Hazel (*Corylus avellana* L.) is a shrub that grows along forest margins, and hazel nuts are a critical fall resource for boar, deer, and squirrels. The nuts ripen during the late summer and drop to the ground in mid-September. Nuts could be picked from the trees, and they could be collected from the ground; but their availability would probably conflict with final harvesting activities. Hazel nuts are easily stored but must be dried before storage. In comparison to other nut bearing species, surprisingly little has been written on the productivity of wild hazel.

Keene (1981:70), however, does estimate its productivity. Assuming squirrels and other wildlife consume 50% of the nuts, he suggests that an average yield of 0.18 kg of whole nuts per plant would be available for human consumption. Keene's estimates are for North America, and they do not include calculations to allow for the voracious feeding of wild boar. In all likelihood, wild boar would consume a significant proportion of the nuts that fell to the ground, further reducing their availability for human consumption. Assuming that wild boar consume half of the available nuts, the net yield declines to 0.09 kg of nuts per plant. Keene suggests a density of 1116 plants per ha, in the optimal areas for hazel growth. Of these, only 30% will produce nuts in any given year, so an average of 30.13 kg/ha can be expected. If the optimal area for hazel growth extends 15 meters into a forest margin, a one kilometer stretch of forest margin would provide 45.19 kg of hazel nuts.

Berries

A variety of raspberries, strawberries, elderberries, and viburnums are available primarily along the forest margins or in clearings. Keene (1981:80) suggests that collectively, 108 kg/ha would be available for human consumption. Assuming that humans actually collected 50% of this amount, then each hectare would produce an average of 54 kg of berries. Since berries are primarily an edge species, a five-meter-wide stretch along one kilometer of forest margin would provide 27 kg of berries.

Tubers

Wild onions and wild garlic are present in the deciduous forest, but their caloric content is low in comparison to those of plants found in marshy areas, cattail roots (*Typha latifolia*), arrowhead (*Sagittaria sagittifolia*), and bulrush (*Scirpus lacustris*). Although the forest tubers were probably consumed, they would have served as a condiment rather than as a staple foodstuff. This discussion concerns the relatively nutritious tubers found in marshy areas. Cattails and bulrushes form dense stands along lake margins, in ditches, and in slow-moving streams. Arrowhead is a component of these marshy plant communities. These tubers are starchy, and their caloric contents can be quite high. Cattail roots, for example, have 3670 calories per kilogram when dried and converted to a flour, and arrowhead root has a caloric value of 1230 calories per kilogram (Asch and Asch 1978:302). Keene (1981:85) estimates that in marshy areas an average of 267.7 kg/ha of tubers would be available. Assuming that 50% of this is used, 133.85 kg tubers could be obtained per ha. A one-kilometer stretch of shoreline ringed by a ten meter wide belt of reeds, cattails, and bulrushes would provide 133.85 kg of tubers.

Greens

The leaves of many plants such as dandelions (*Taraxacum* sp.), dock (*Rumex* sp.), goosefoot (*Chenopodium* sp.), and knotweed (*Polygonum* sp.) are edible. Their use could have begun in spring and continued throughout the summer; but the leaves of these plants tend to become bitter as the plant matures and produces seeds. Therefore, the consumption of greens probably would have tapered off through the summer. Keene (1981:87) estimates that throughout the summer, one hectare would produce 12.1 kg of edible greens.

Seeds

Seeds from wild grasses (Gramineae), sedges (*Carex* sp.), bulrushes, water lilies (*Nuphar lutea* L.), dock, goosefoot, and knotweeds could have been consumed. These plants grow in a variety of environments. Dock, goosefoot, and knotweeds are pioneer annuals and perennials. Their abundance in the natural forest would be limited, although their density would increase with the appearance of farming communities. Their use may have been minimal during the Mesolithic period, but it probably increased after the introduction of farming. Wild grasses and sedges are common along lakeshores, and their densities would have been highest in these areas. Bulrushes are characteristic of shallow water communities, and water lilies can tolerate water several meters deep. Simms (1984) provides figures that suggest 1175 kg/ha of clean seed can be obtained. This figure is high, and intuitively it seems improbable for Europe. One third of the figure, 387 kg of clean seed, has arbitrarily been adopted as the average seed yield.

Modeling Monthly Resource Use

Jochim's model (1976) calls for establishing the proportional contribution of each wild resource to the *annual* diet and then determining the *monthly* use of each. Jochim's model is briefly presented below. Readers wishing an exhaustive discussion are referred to his original monograph.

Proportional Contributions to the Annual Diet

The mobility, density, aggregation, non-food value, and size of red deer, roe deer, beaver, fish, and small game interact to affect the relative importance of each within the total proportion of wild game in the annual diet. Changes in any values of these variables for any wild resource influence the relative importance of *all* resources. This is critical in evaluating the effects that the activities of the farmers may have had on the availability of game.

As discussed earlier, red deer are direct competitors with livestock, and red deer densities near villages may have been low. Jochim's (1976: Table 20) calculations using the normal red deer density of four per square kilometer were adopted for one set of analyses. As Table 50 Column 2 shows, under a normal red deer density of four individuals per square kilometer, red deer constitute 32.9% of the wild game, roe deer 4.0%, boar 28.3%, fish

Table 50

Relative Importance of Game
(% of total wild game)

Resource	Red Deer Density	
	Normal†	Low
Red Deer	32.9	30.5
Roe Deer	4.0	4.0
Boar	28.3	29.0
Fish	16.8	18.0
Beaver	1.5	1.5
Small Game	16.5	17.0

†Variations in calculations presented here and by Jochim (1976:107) result from the effects of rounding.

16.8%, beaver 1.5%, and small game 16.5%. To judge the effects that the decreased availability of red deer would have had on hunting strategies, the red deer density was lowered to two animals per square kilometer, and the relative importance of each class of game was recalculated (Table 50, Column 3). The significance of red deer decreases by 2.4% while that of beaver and roe remains constant. The relative importance of boar, fish, and small game rises 0.7%, 1.2%, and 0.5% respectively.

The significance of each class of game in the total diet is calculated by multiplying its relative importance by the proportion of wild game in the annual diet. Assume for the moment that wild game provides 100% of the annual diet. Multiplying the relative importance of each class of resources by 1.0, red deer provide 32.9%, roe deer 4.0%, boar 28.3%, fish 16.8%, beaver 1.5%, and small game 16.5% of the annual diet. Unfortunately, wild game do not provide 100% of the diet: the amount of wild game in the diet is determined by the quantity of domestic resources, wild plants, and birds in the annual diet.

Jochim (1976:107) suggested that wild plant foods provided 20% and birds 2% of the forager's diet. Returning to the issue of estimating the annual proportion of each class of game, if the

Table 51

Annual Proportions of Resource Classes

Strategy	Domestic			Wild	
	Crops	Meat	Milk	Plants	Game
Farmers					
Optimal Diet	62.9	15.3	7.7	2.8	11.3
Foragers					
No Domestic Resources	0.0	0.0	0.0	20.0	80.0

foragers do not receive wheat from the farmers, plants make up 20% of the forager diet, and birds 2%; all other wild game account for 78% (Table 51). Assuming a normal red deer density, then red deer would provide 32.9% of the 78%. This amounts to 25.7% of the annual diet. The proportions of all other classes of game are determined similarly by multiplying their relative importance by 78%.

Calculations for the farmers are made somewhat more complicated. Birds and small game were lumped together for the farmers, and the contribution of wild plants was set at 20.0% of the total proportion of wild resources in the annual diet. The optimal model presented in the previous chapter (Table 43) suggests crops supplied 62.9% of the annual diet, domestic meat 15.3%, and milk 7.7%. Wild resources, therefore, supplied 14.1%. Plants account for 20% of wild resources. This constitutes 2.8% of the annual diet. Subtracting 2.8 from 14.1, we find that wild game furnish the remaining 11.3%. As above, the annual proportion of each class of wild game is obtained by multiplying its relative importance by the proportion of wild game in the annual diet. Thus, under normal red deer densities (Table 50, Column 2), red deer make up 32.9% of the 11.3%. This is the equivalent of 3.7% of the *annual* diet under the optimal Neolithic farming strategy (Table 52). The proportion of the diet fulfilled by roe deer, boar, beaver, fish, small game, and birds also calculated by multiplying

Table 52

Annual Proportions of Each Game Resource:
Normal Red Deer Densities

Strategy	Wild Game Resource					
	Red Deer	Roe Deer	Boar	Fish	Beaver	Birds & Small Game
Farmers						
Optimal Diet	3.7	0.4	3.2	1.9	0.2	1.9
Foragers						
No Domestic Resources	25.7	3.0	22.1	13.1	1.2	14.9

their relative importance by 11.3%. The relative importance of each resource is used along with the monthly resource fraction to determine its monthly use.

Monthly Resource Fractions

The second step in modeling wild resource exploitation is to determine the monthly use of each wild resource. This is accomplished by calculating the proportion of each resource consumed in any given month. The resulting figure is called the monthly resource fraction. If a resource is consumed equally throughout the year, then 1/12th (8.3%) of the annual allocation is consumed each month. Fluctuations depend on the seasonal abundance and quality of a resource, as well as a population's willingness and ability to store the resource. The monthly resource fractions for small game, red deer, roe deer, boar, fish, and beaver (Table 53) were adopted directly from Jochim (1976:107–117). The figures for wild plant foods and fowl were modified slightly for the farmers; moreover, domestic resources were added for both the foragers and the farmers. These changes are discussed next.

Table 53

Monthly Resource Fractions
(% of annual total)

Resource	Month (September – August)											
	Sept	Oct	Nov	Dec	Jan	Feb	Mar	Apr	May	June	July	Aug
Red Deer	8.5	8.0	6.5	7.5	15.0	14.0	15.0	6.5	4.0	3.5	5.5	6.0
Roe Deer	7.5	7.5	7.5	9.5	12.5	11.5	12.0	8.5	6.0	5.5	5.5	6.5
Boar	7.5	7.5	13.5	13.5	12.0	10.0	10.0	4.0	4.0	5.5	6.0	6.5
Fish	14.0	3.0	2.0	2.0	1.5	2.0	3.5	7.0	25.0	14.0	13.0	13.0
Beaver	0.0	0.0	0.0	20.0	20.0	20.0	20.0	20.0	0.0	0.0	0.0	0.0
Birds	0.0	25.0	25.0	0.0	0.0	0.0	25.0	25.0	0.0	0.0	0.0	0.0

Source: Jochim 1976:107–117

Table 54

Farmer Modifications to Monthly Resource Fractions:
Crops and Wild Plants
(% of annual total)

Resource	Month (September-August)											
	S	O	N	D	J	F	M	A	M	J	J	A
Wild Plants	12.0	10.0	7.0	5.0	5.0	5.0	5.0	7.0	10.0	10.0	12.0	12.0
Crops	8.2	8.2	8.2	8.2	8.2	8.2	8.6	8.6	8.2	8.2	8.6	8.6

Wild plant foods

Jochim assumed the foragers consumed plant foods from April through November. This assumption can be justified for mobile hunter-gatherers, if it is assumed they had little or no storage. The assumption is not warranted for the farmers; thus the resource fractions for wild plant foods were modified to allow for usage throughout the year (Table 54). Plant foods are seasonally abundant. Tubers are available throughout the year; however, their quality is best in the early spring—before growth saps the tubers of their stores of starches. Greens appear in early spring, but their palatability declines as the summer passes. Seeds and berries generally become available in late June, and their abundance increases throughout the summer. Finally, hazel nuts ripen in mid-September. Since there is so little understanding of the significance of plant foods in the Early Neolithic diet, it is arbitrarily assumed that tubers would provide 95% of the spring wild plant foods and greens the remaining 5%. Tubers, greens, seeds, and berries have respectively been assigned 30%, 5%, 50%, and 15% of the summer portion of wild plant foods. In the fall, seeds comprise 20% and hazel nuts 80% of the plant foods. Finally, hazel provides 80% and seeds 20% of the winter wild plant requirements (Table 55).

Table 55

Farmer Seasonal Consumption of Wild Plant Foods
(expressed as % of wild plant foods each season)

Resource	Fall	Winter	Spring	Summer
Tubers	0	0	95	30
Greens	0	0	5	5
Seeds	20	20	0	50
Berries	5	0	0	15
Hazel nuts	75	80	0	0

Wild Fowl

As discussed above, wild fowl were classed with small game the farmers' strategy was evaluated.

Domestic Resources

For the farmers it was assumed that wheat and lentils would be eaten throughout the year, but their monthly fractional importance must be adjusted to allow for heavy labor during planting and harvesting seasons. For eight months of the year, crops accounted for 8.2% of the annual diet. This was increased to 8.6% during spring planting in March and April and for the harvest in July and August (Table 54). The fractional importance of both meat and milk were also adjusted to reflect their seasonal availability (Table 56). Meat is considered first.

The death rate of calves, kids, and lambs was used to determine the number of individuals that would have died from birth to weaning. For both lambs and kids, it was assumed that 10% of the deaths occurred in February, at the end of winter and during the early stages of lambing and kidding; 25% in March; and 15% in April. The remaining lamb and kid deaths were spread evenly throughout their respective milking periods. It was also assumed that 50% of the ewe and doe losses occurred from February through the end of the milking period. The rest were spread evenly throughout the year. Similarly, the calf deaths and half of the

Table 56

Farmer Modifications to Monthly Resource Fractions:
Cattle, Ovicaprids, and Pigs
(% of resource annual total)

Resource	Month (September-August)											
	S	O	N	D	J	F	M	A	M	J	J	A
Live-stock	5.7	7.2	7.2	7.2	7.2	12.4	14.5	14.0	7.3	5.9	5.7	5.7
Milk	9.0	0.0	0.0	0.0	0.0	0.0	16.0	16.0	16.0	16.0	14.0	13.0

cow losses were assumed to have occurred from calving to the end of milking; the remaining were spread evenly throughout the year. As discussed in Chapter 5, pigs are assumed to have been slaughtered in the fall. Their total contribution to the diet was divided into the four-month period from October through January.

Sheep produce milk for 135 days, cattle for 200 days, and goats for 210 days. Assuming milk became available for human consumption in March, milk from all three species would have been available until mid-July; cow and goat milk would have been available from mid-July until early September and goat milk until late September. To complicate matters, the quantity and nutritional value of sheep, cow, and goat milk are not the same. The monthly fraction of milk was obtained by converting the daily yield of each species to calories. These figures were summed for the 135 days on which all three were producing. Cow and goat yields were totaled for the 65 days on which these two species were producing. Finally, the daily yield of goats was determined for 10 days. The monthly fractions were obtained by summing the daily calories and determining each month's proportion of the annual caloric yield of milk.

One set of analyses called for examining a diet consisting of ovicaprid, but not cattle products. In this case, resource fractions for meat and milk were determined for resources from sheep, goats, and pigs (Table 57).

In order to establish resource fractions for the domestic foods in the forager diet, it was necessary to determine when and in

Table 57

Farmer Modifications to Monthly Resource Fractions: No Cattle
(% of resource annual total)

Resource	Month (September-August)											
	S	O	N	D	J	F	M	A	M	J	J	A
Live-stock	3.5	7.3	7.3	7.3	7.3	9.4	21.1	13.2	6.7	6.7	6.7	3.5
Milk	8.0	0.0	0.0	0.0	0.0	0.0	18.0	18.0	18.0	18.0	12.0	8.0

what quantity domestic resources would have been added to the diet (Table 58). Chapter 6 suggests forager-farmer interaction may have been as brief as a few weeks in the late winter/early spring. From the forager's perspective, this comes before floral resources are readily available and at a time when game is both hard to find and of poor nutritional quality (Speth and Spielmann 1982). Domestic resources would have been welcome at this time. An important aspect of hunter-gatherer subsistence is the forager's ability to monitor and predict resource availability. Harvest quality would certainly affect the amount of grain that the farmers .would be able to provide at the end of the winter. It would have been important for foragers to evaluate the harvest quality each fall to determine the quantity of resources that would be available the following spring. It was estimated that the foragers would have consumed 20% in the late fall/early winter, and 80% in the spring. An allocation was also made for meat received from the farmers. Heavy livestock losses will follow the birth of calves, lambs, and kids. If this meat exceeds the amounts that the farmers can use, it will be available for trade to the farmers. It is assumed that 60% of this meat will be consumed in March and 40% in April. The figures shown in Table 58 were added to Jochim's monthly resource fractions in Table 53.

Table 58

Forager Modifications in Monthly Resource Fractions
(% of resource annual total)

Resource	Month (September-August)											
	S	O	N	D	J	F	M	A	M	J	J	A
Live-stock	0.0	0.0	0.0	0.0	0.0	0.0	60.0	40.0	0.0	0.0	0.0	0.0
Wheat	0.0	0.0	10.0	10.0	0.0	0.0	40.0	40.0	0.0	0.0	0.0	0.0

Monthly Diet

Having determined the proportional contribution of each resource to the annual diet and the monthly resource fraction, the monthly diets can be estimated. Values for monthly diets are obtained by multiplying the monthly fraction of each resource by its proportion of the annual diet. For the foragers, one-twelfth (8.3%) of the annual diet must be obtained each month. In theory, resources can be used as they became available. But as seasonality increases one or more resources must be used as a buffer to ensure that one-twelfth of the annual requirements are obtained each month. Jochim used small game as the buffering resource to fulfill the dietary requirements each month. To obtain the percentage of small game in the diet, the contribution of all other resources is totaled and subtracted from 8.3%. Small game is allocated the amount required to obtain 8.3% of the annual diet.

This approach worked for Jochim's application of the model, because seasonality involved resources that provided a very small proportion the annual diet. Difficulties were encountered when domestic resources were added to the forager diet, and modifications were needed. Wheat may have fulfilled 3.8%, 7.7%, or 11.5% of the forager diet, and domestic meat or milk may have been available. By assuming 80% of the wheat and 100% of the livestock resources were consumed in a two-month period in the late winter/early spring, the monthly total could exceed one-twelfth the annual diet. Small game was used as a buffer to fill in for any deficits; however, resource exploitation must be reduced in those months during which a dietary surplus occurs.

Slightly more complicated problems were encountered when evaluating the farmer diet. Heavy labor was required in both the planting (March/April) and the harvesting (July/August) seasons, and this increased the amount of calories needed during these periods. In all, 34.4% of the annual diet was needed during these four months. This amounts to 8.6% of the diet per month during planting and harvest seasons. The remaining 65.6% of the annual diet was divided evenly between the other eight months. The monthly proportion of the annual diet in each of these months was 8.2%.

Milk and meat availability in the Neolithic diet caused particular problems. Milk provided up to 7.7% of the Neolithic diet, but its consumption was limited to a seven-month period. When cows, ewes, and does provided milk, 16% of the annual yield was produced during each of the first four months milk was available; 14% and 13% respectively was produced during the fifth and sixth months, and 9% during the seventh month. The problem was exacerbated when only ovicaprid milk was modeled, for 18% was produced each month for the first four months, 12% the fifth month, and 8% during both the sixth and seventh months. The availability of domestic meat also fluctuated monthly, the largest quantities becoming available in March and April, when milk was also available in relatively large quantities. Livestock resources in March and April caused consumption to exceed the farmers' requirements in some strategies.

Additional buffering resources were needed to ensure the monthly proportion of the annual diet was attained but not exceeded. Therefore, the monthly proportions of wild plants, boar, red deer, fish, and roe deer were reduced in that order. In determining the structure of the diet, the monthly proportions were totaled. If the sum was *less* than 8.3% of the annual total in the forager diet (or either 8.6% of the annual farmer diet during planting and harvesting or 8.2% of the farmer diet in the remaining eight months), then the difference was allotted to the monthly proportion of small game in the diet. If the total *exceeded* the requisite proportion, then one-half the difference was subtracted from the wild plant proportion and the remainder was subtracted from the boar proportion. Remaining surpluses were then subtracted from the red deer allocation followed by the fish and the roe deer.

Once the monthly proportions are determined, a resource use schedule is obtained by plotting the monthly proportions of each resource.

Hunting and Gathering in a Farming Economy

Archaeologists have not yet determined whether or in what degree Neolithic farmers incorporated domestic meat and milk in their diets. In order to evaluate the effects that the consumption of meats and milks had on Neolithic hunting strategies, calculations were made for diets with four different resource configurations but with wheat providing 62.9% of each diet. The four strategies are: (1) the "optimal" strategy, including 40 cattle, 40 ovicaprids, and 2 sows; (2) a "no milk" strategy that incorporates only meat from the livestock; (3) a "no livestock" strategy, including only domestic crops and wild resources; and (4) an "ovicaprid" strategy that consists of crops as well as meat from two litters of pigs and resources from an ovicaprid herd of 50 sheep and 50 goats. Calculations were made for wild resource exploitation assuming a normal red deer density (Strategies A–D, Table 59) and a low red deer density (Strategies E–H, Table 59). In all, eight Neolithic strategies were examined.

Wild Resource Exploitation with Normal Red Deer Densities

As discussed in the previous section, the annual proportion of each resource was obtained by multiplying its relative importance (Table 50 Column 2) by the annual proportion of wild game in each subsistence strategy (Table 59: A–D) to obtain the annual proportion of each resource (Table 60: A–D). The annual proportion of each resource was then multiplied by the monthly resource fractions (Table 61—and in the case of the Strategy D, Table 62) to obtain the monthly diets (Appendix: A-D). Resource use schedules are presented in Figure 6 and the exploitation patterns of red deer, boar, small game, and fish appear in Figure 7.

The calendar year has been divided into seasons corresponding to the agricultural cycle beginning in September, following the

Table 59

Farmer Subsistence Strategies
(% of annual diet)

Strategy	Domestic			Wild	
	Crops	Live-stock	Milk	Plants	Game
Normal Red Deer Densities					
A. Optimal	62.9	15.3	7.7	2.8	11.3
B. No milk	62.9	15.3	0.0	4.4	17.4
C. No livestock	62.9	0.0	0.0	7.4	29.7
D. Ovicaprids	62.9	3.1	6.0	5.6	22.4
Low Red Deer Densities					
E. Optimal	62.9	15.3	7.7	2.8	11.3
F. No milk	62.9	15.3	0.0	4.4	17.4
G. No livestock	62.9	0.0	0.0	7.4	29.7
H. Ovicaprids	62.9	3.1	6.0	5.6	22.4

Table 60

Farmer Resource Proportions
(% of annual diet)

Resource	Normal Red Deer Densities				Low Red Deer Densities			
	Optimal (A)	No Milk (B)	No Livestock (C)	Ovicaprids (D)	Optimal (E)	No Milk (F)	No Livestock (G)	Ovicaprids (H)
Red Deer	3.7	5.7	9.8	8.0	3.4	5.3	9.1	7.4
Roe Deer	0.4	0.7	1.2	1.0	0.5	0.7	1.2	1.0
Boar	3.2	4.9	8.4	6.9	3.3	5.0	8.6	7.1
Fish	1.9	2.9	5.0	4.1	2.0	3.1	5.3	4.4
Beaver	0.2	0.3	0.4	0.4	0.2	0.3	0.5	0.4
Small Game	1.9	2.9	4.9	4.0	1.9	3.0	5.0	4.1
Plants	2.8	4.4	7.4	6.7	2.8	4.4	7.4	6.7
Livestock	15.3	15.3	0.0	2.4	15.3	15.3	0.0	2.4
Milk	7.7	0.0	0.0	3.6	7.7	0.0	0.0	3.6
Crops	62.9	62.9	62.9	62.9	62.9	62.9	62.9	62.9

wheat harvest. Fall includes September, October, and November; winter lasts from December through February; spring begins in March and continues through May; and summer consists of June, July, and August. The structure of each subsistence strategy will be examined.

A. Optimal Strategy

In the optimal strategy presented in Chapter 6, domestic resources provide 85.9% of the diet, wild plant foods provide 2.8%, and wild game fulfills the remaining 11.3% . While small game and birds are important resources in the fall and winter, every other class of wild resources provides less than 8% of the monthly diet. Although they are not of major importance, boar, red deer, and fish do show some seasonality, and the farmers could be expected to hunt these species in the fall, winter, and spring respec-

Table 61

Farmer Monthly Resource Fractions: All Livestock
(% of annual total)

Resource	Sept	Oct	Nov	Dec	Jan	Feb	March	April	May	June	July	Aug
Red Deer	8.5	8.0	6.5	7.5	15.0	14.0	15.0	6.5	4.0	3.5	5.5	6.0
Roe Deer	7.5	7.5	7.5	9.5	12.5	11.5	12.0	8.5	6.0	5.5	5.5	6.5
Boar	7.5	7.5	13.5	13.5	12.0	10.0	10.0	4.0	4.0	5.5	6.0	6.5
Fish	14.0	3.0	2.0	2.0	1.5	2.0	3.5	7.0	25.0	14.0	13.0	13.0
Beaver	0.0	0.0	0.0	20.0	20.0	20.0	20.0	20.0	0.0	0.0	0.0	0.0
Wild Plants	12.0	10.0	7.0	5.0	5.0	5.0	5.0	7.0	10.0	10.0	12.0	12.0
Livestock	5.7	7.2	7.2	7.2	7.2	12.4	14.5	14.0	7.3	5.9	5.7	5.7
Milk	9.0	0.0	0.0	0.0	0.0	0.0	16.0	16.0	16.0	16.0	14.0	13.0
Crops	8.2	8.2	8.2	8.2	8.2	8.2	8.6	8.6	8.2	8.2	8.6	8.6

Table 62

Farmer Monthly Resource Fractions: Pigs, Goats, and Sheep
(% of annual total)

Resource	Sept	Oct	Nov	Dec	Jan.	Feb.	March	April	May	June	July	Aug.
Red Deer	8.5	8.0	6.5	7.5	15.0	14.0	15.0	6.5	4.0	3.5	5.5	6.0
Roe Deer	7.5	7.5	7.5	9.5	12.5	11.5	12.0	8.5	6.0	5.5	5.5	6.5
Boar	7.5	7.5	13.5	13.5	12.0	10.0	10.0	4.0	4.0	5.5	6.0	6.5
Fish	14.0	3.0	2.0	2.0	1.5	2.0	3.5	7.0	25.0	14.0	13.0	13.0
Beaver	0.0	0.0	0.0	20.0	20.0	20.0	20.0	20.0	0.0	0.0	0.0	0.0
Wild Plants	12.0	10.0	7.0	5.0	5.0	5.0	5.0	7.0	10.0	10.0	12.0	12.0
Livestock	3.5	7.3	7.3	7.3	7.3	9.4	21.1	13.2	6.7	6.7	6.7	3.5
Milk	8.0	0.0	0.0	0.0	0.0	0.0	18.0	18.0	18.0	18.0	12.0	8.0
Crops	8.2	8.2	8.2	8.2	8.2	8.2	8.6	8.6	8.2	8.2	8.6	8.6

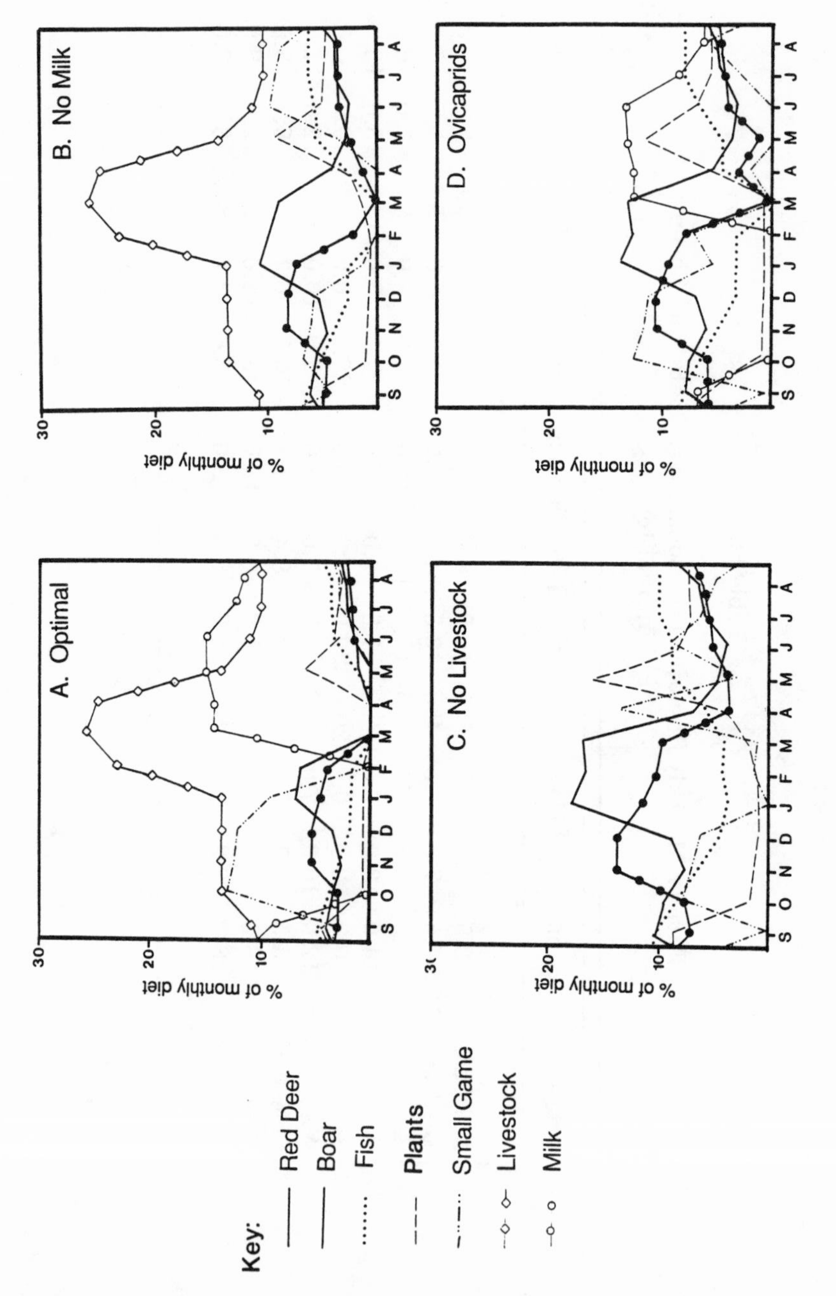

Figure 6. Farmer Resource Use: Strategies A-D

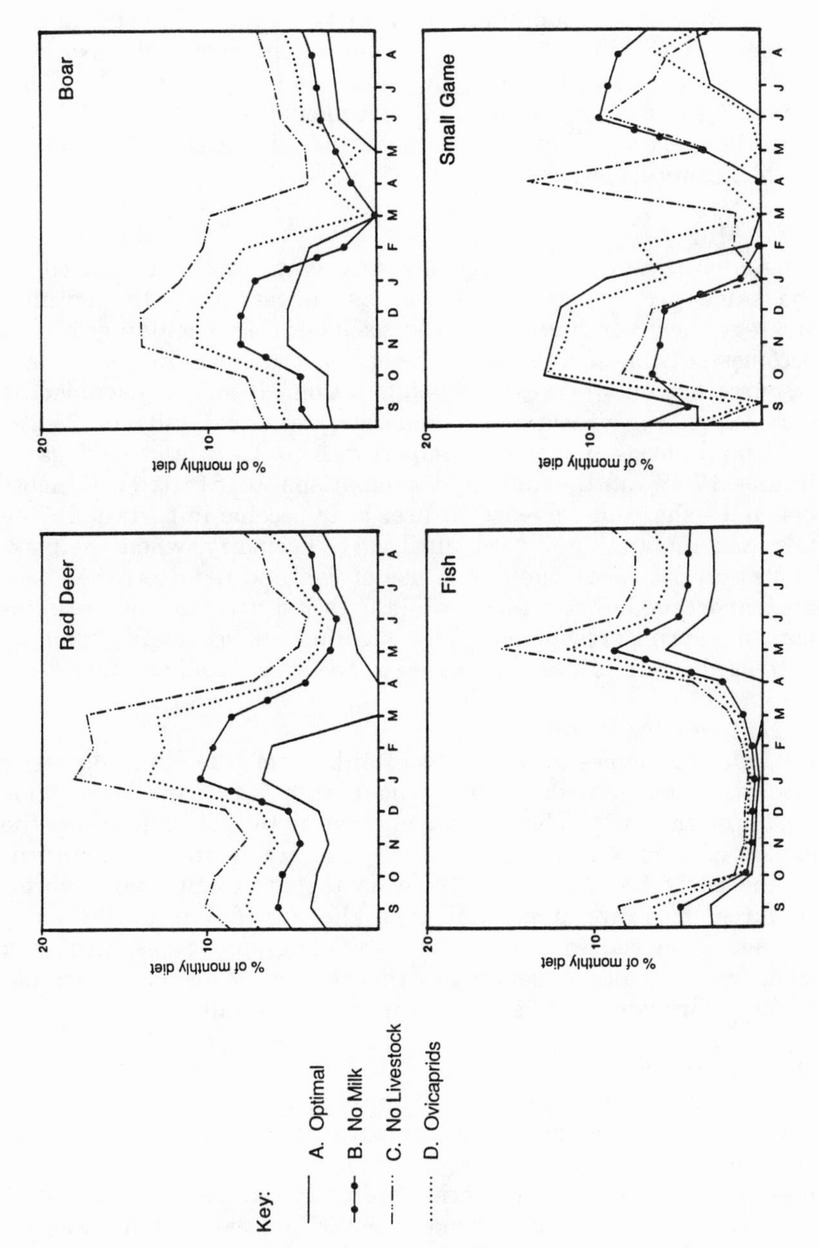

Figure 7. Farmer Exploitation Patterns: Strategies A-D

tively. Milk is the dominant resource throughout the spring and summer; it continues to be important in the early fall. No wild resources are needed during the months of March and April because of the availability of milk and meat from the livestock. In fact, there is a small surplus of meat and milk that could be given to the hunter-gatherers.

B. No Milk

As with the first strategy, domestic crops and meat constitute the same percentage of the diet as suggested by the optimal strategy; however, milk has been excluded. By excluding milk it becomes possible to determine the effects that including milk as a resource has on wild resource exploitation. If milk is excluded as a resource, the proportion of domestic resources falls to 78.2%; wild plant foods increase in importance to 4.4%, and wild game fulfills 17.4% of the diet. In comparison to Strategy A, most resources show an increase in use: boar become important in the fall; and remain significant until early February, when their exploitation decreases due to the use of fish and red deer. Red deer are important in the winter; while both red deer and fish are important spring resources. The significance of small game is decreased in the fall but increased in the spring and summer.

C. No Livestock Products

Deleting domestic meat and milk from the diet, domestic resources now provide 62.9%, wild plants 7.4%, and wild game 29.7% of the diet. These changes have noticeable effects on the exploitation of wild resources. Red deer are relatively important in the early fall, and they serve as the most important winter resource; boar are important from the late fall until the early spring. Fish continue to be a significant spring resource, and small game is also important in this season. Wild plants are important throughout the summer into the early fall.

D. Ovicaprid Resources

If a herd of 50 goats and 50 sheep is raised instead of cattle, and their resources are incorporated in the subsistence, then the structure of wild resource exploitation changes again. Meat and milk from a combined ovicaprid herd of 100 animals provide 9.0% of the diet. With crops providing 62.9% of the annual calories, wild plants contribute 5.6%, and wild game 22.5% of the diet. The importance of red deer drops significantly in both fall and winter. Small game once again becomes an important fall resource but its significance in the spring is diminished considerably. Boar first become important in the late fall and is significant only through mid-winter. Fish continue to be an important spring resource, and wild plants remain important in the summer and early fall.

Comparative Efficiency of Strategies A–D

Plants, boar, and small game are most seriously affected by the addition and subtraction of domestic resources to the diet. Variation in the importance of either boar or plants in the spring or summer could be attributed to the abundance of milk in these seasons. This would affect the first and fourth strategy primarily during the months of March, April, and May. Indeed, fluctuations in the dietary proportion of boar are noticeable at this time. The role of small game is critical. Small game is assumed to be a buffering resource. The proportion of small game increases or decreases as a result of net changes in the proportion of all other wild game in the diet. Small game is significant during the fall and winter in the strategies including milk; but in other strategies, small game is more important during the spring and summer.

The use of proportions is somewhat misleading, for they provide little feeling for the quantity of game that must be hunted or trapped. Hunting requirements can be calculated by converting the monthly proportion of each resource into calories, and then dividing that figure by the number of calories the average individual of the particular species would provide. Jochim (1976:134) suggested that red deer would provide an average of 217,000 kilocalories; roe deer, 34,000; boar, 378,000; beaver, 56,000; fish, 650; and small game 3,888. Summing the monthly requirements, the annual game requirements are obtained (Table 63). Assuming that the game is harvested at 20% of its density, the minimum required hunting territories can be estimated (Table 64).

Although we have little concrete information on the exploitation of wild plant foods, there is no doubt that greens, nuts, berries, tubers, and seeds would have been exploited. It has been assumed that 20% of the total wild resources in the diet will be fulfilled by plant foods. In order to obtain a better understanding of

Table 63

Farmer Game Requirements: Normal Red Deer Densities
(expressed as number of individuals)

Strategy	Red Deer	Roe Deer	Boar	Fish†	Bea- ver	Small Game
(A) Optimal	3.5	2.8	1.9	724.5	0.5	314.8
(B) No milk	7.2	5.6	2.9	1269.9	1.3	318.6
(C) No livestock	12.6	9.5	6.1	2167.5	2.3	379.9
(D) Ovicaprids	9.5	7.2	4.0	1642.1	1.7	343.7

†Fish are assumed to average one kilogram per individual

Table 64

Minimum Territory Needed for Each Resource:
Normal Red Deer Densities
(km^2)

Strategy	Red Deer	Roe Deer	Boar	Fish	Bea- ver	Small Game
(A) Optimal	4.4	1.2	0.8	37.0	3.7	15.2
(B) No milk	9.0	2.3	1.2	64.8	10.2	15.5
(C) No livestock	15.7	4.0	2.5	110.6	18.0	18.4
(D) Ovicaprids	11.9	3.0	1.7	83.8	13.3	16.7

the potential contribution of wild plants to the diet, it was assumed tubers and greens would have been eaten only in the spring and summer, that seeds and berries would have been consumed throughout the summer, and that seeds, berries, and nuts would have been stored for consumption throughout the winter. These assumptions may ultimately prove unrealistic; nonetheless, they provide a starting point for the analysis.

In order to estimate the quantities of plant foods needed in each season, the seasonal proportion of each plant food was converted into calories. This figure was then divided by the number of calories provided by a kilogram of the resource to obtain the amount of plant food needed. Table 65 shows the requisite amount of each resource for each of the four subsistence strategies. Table 66 estimates the area needed to fulfill these requirements. These plant foods are not a major component of the farmers' diet, and a relatively small area of shoreline would be needed to fulfill the tuber requirements; the berries and hazel could be collected from forest margins surrounding fields; greens would have to be collected from areas away from the village within the unused browsing area.

Table 65

Farmer Annual Requirements of Wild Plant Foods: Normal Red Deer Densities (kg)

Resource	Strategy			
	Optimal (A)	No Milk (B)	No Live-stock (C)	Ovi-caprids (D)
Tubers	33.3	95.2	205.8	117.0
Greens	60.8	124.7	240.5	155.8
Seeds	74.1	112.5	194.0	148.3
Berries	77.1	141.8	169.1	70.2
Hazel nuts	98.29	133.2	251.6	196.6

Table 66

Area Needed for Neolithic Plant Exploitation:
Normal Red Deer Densities

Resource	Strategy			
	Optimal (A)	No Milk (B)	No Livestock (C)	Ovicaprids (D)
Tubers (km shoreline)	0.25	0.71	1.54	0.87
Greens (ha)	5.02	10.31	19.88	12.88
Seeds (ha)	0.19	0.29	0.50	0.38
Berries (km forest edge)	2.86	5.25	6.26	2.60
Hazel nuts (km forest edge)	2.18	2.95	5.57	4.35

Although red deer and boar are dominant wild resources, they are not the limiting factor in determining minimal territory sizes. Fish are. Fish requirements call for a territory of from 37.0 to 110.6 square kilometers, depending on the amount of domestic resources in the annual diet. Farmers who incorporate the use of both ovicaprids and cattle (Strategy A) require a territory that is 33.4% the size of farmers who use no livestock products (Strateagy C); and 44.2% the size required by farmers who incorporate only ovicaprid resources (Strategy D) in their diet. Finally, villages that consume domestic meat, but no milk (Strategy B) require a territory that is 58.6% of the size needed by villagers who use no domestic livestock products (Strategy C).

Of particular interest are the effects that farming activities may have had on wild game availability. Forest clearance may have reduced the habitat of roe deer, boar, and small game. On the other hand, it would have resulted in a larger expanse of forest margin, the environment in which these resources thrive. Thus farming activities may actually enhance the environment for these resources, and it cannot automatically be assumed that their densities would decrease. It is unlikely Early Neolithic villages could have affected the environmental quality of the lakes and streams to the extent that they would have reduced net fish density. As discussed earlier, however, red deer compete with cattle and sheep. Cultivating and stockbreeding activities may have resulted in a lowered red deer densities. The following section considers the effects that lowered red deer densities may have had on the structure of the farmers' wild resource exploitation.

Effects of low red deer densities on hunting and gathering strategies

A decrease in the density of any species affects the relative importance of all resources. In this case, a decrease in red deer density reduces the relative proportion of red deer from 32.9% to 30.5% of the total wild game while increasing the significance of fish from 16.8% to 18.0% (Table 50, Column 3). This in turn affects resource use and the pattern of resource exploitation and the number of individuals required annually. The nature and significance of these changes are considered next. The strategies below correspond to the four discussed earlier. As before, each monthly diet (Appendix: E-H) was obtained by multiplying the subsistence strategies (Table 59) by the farmer monthly resource fractions (Tables 61 and 62). The resource use schedule and exploitation patterns are shown in Figures 8 and 9 respectively.

E. Optimal Diet

As with Strategy A above, domestic foodstuffs provide 85.9% of the diet, wild plants fulfill 2.8%, and wild game contribute the remaining 11.3%. The only difference between Strategies A and E is that the latter's relative wild game frequencies are based on a low red deer density. As with Strategy A, small game is an important resource throughout the fall and winter, while milk is important throughout the spring and summer. Boar, red deer, and fish

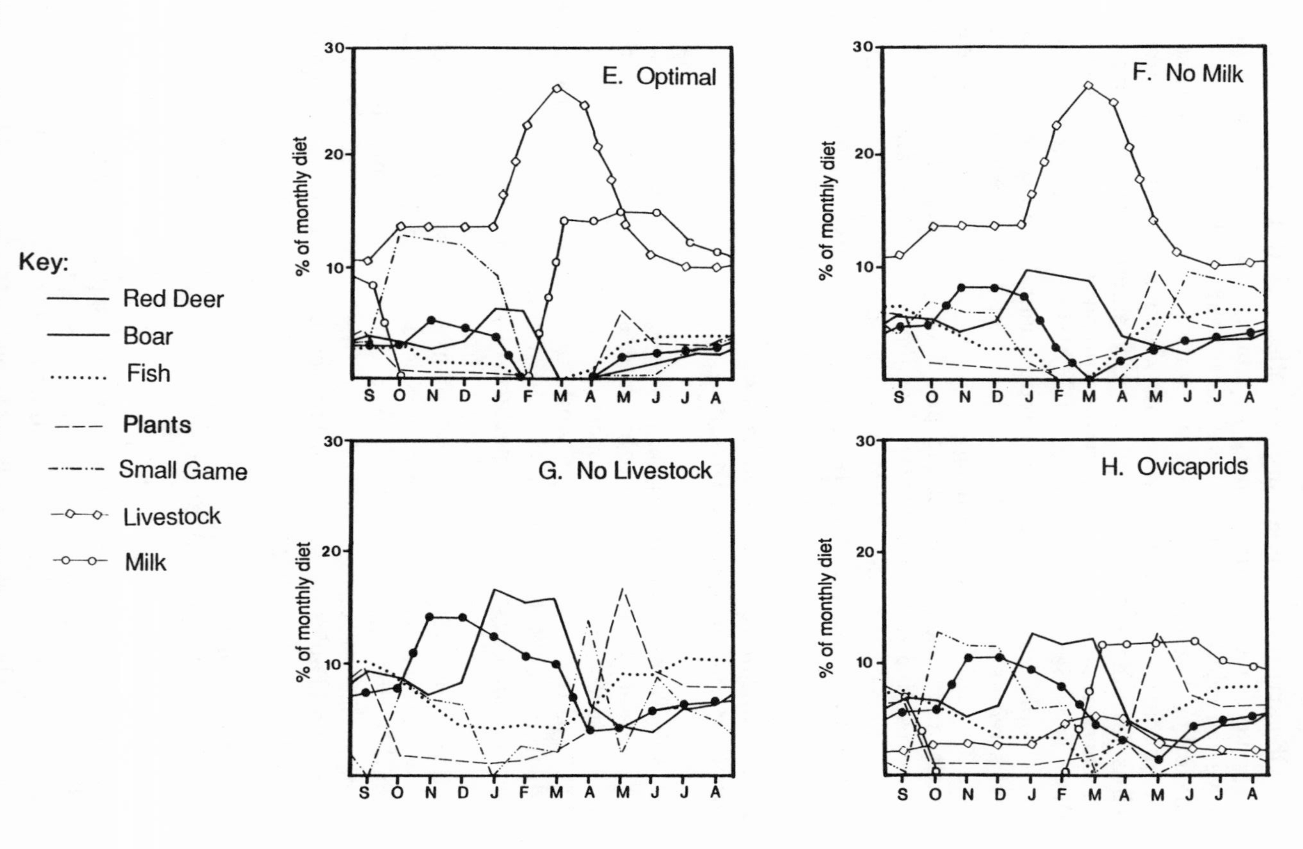

Figure 8. Farmer Resource Use: Strategies E-H

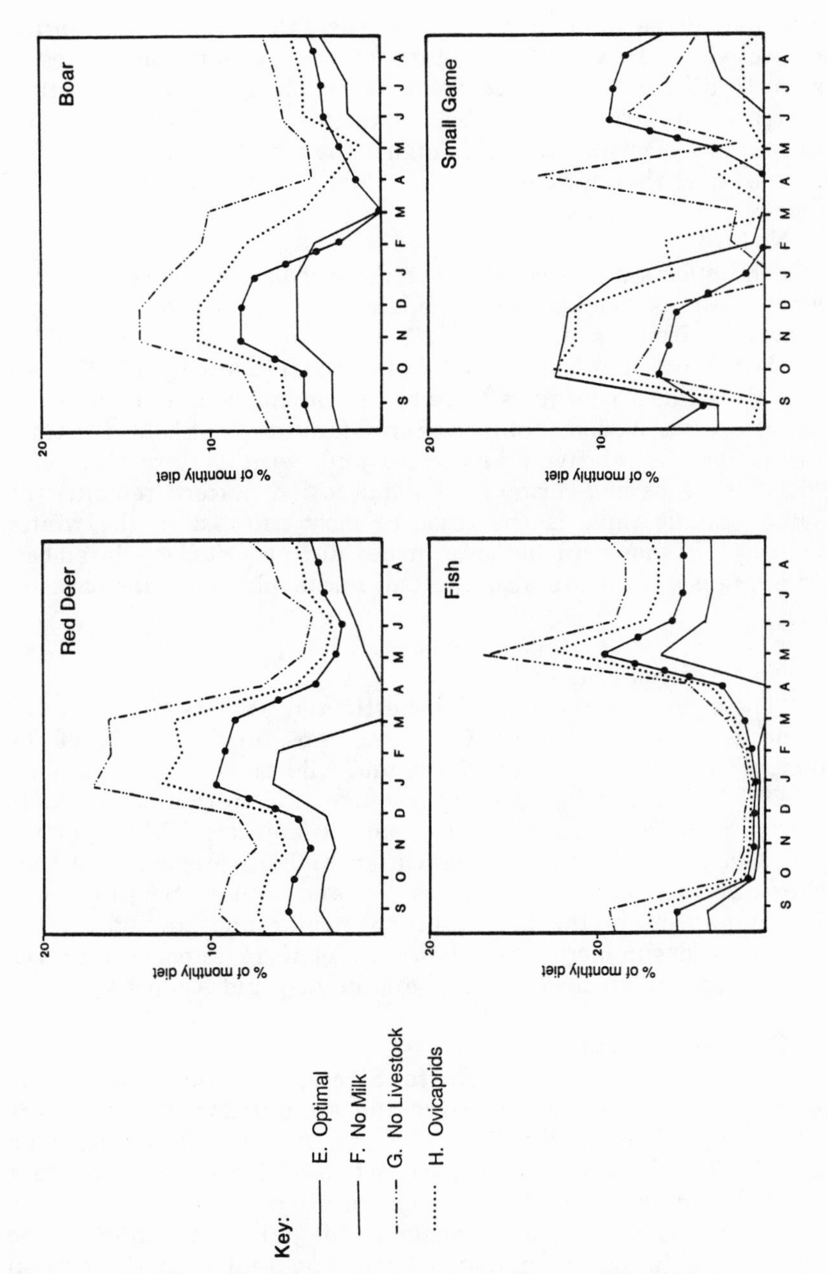

Figure 9. Farmer Exploitation Patterns: Strategies E-H

continue to be more important in the fall, winter, and spring respectively. However, the shape of the exploitation curves is slightly different. Red deer plateau in January and thereafter decrease in significance. Boar exploitation remains virtually unchanged. During July and August the significance of fish is increased over that of Strategy A.

F. No Milk

The effects of a low red deer density become more apparent when milk is deleted from the diet. As in Strategy B above, domestic resources provide 78.2% of the diet, wild plants add a further 4.4%, and game contributes the remaining 17.4%. Although red deer continues to be an important winter resource, its dietary contribution declines from January until May. Once again, boar is relatively unaffected and, with the exception of a slight increase in February, its exploitation pattern remains the same. Small game is the resource most affected in the winter months. It shows an increase in use during October, November, and August. Fish are also affected, and exploitation increases in May.

G. No Livestock Products

The overall proportions of domestic and wild resources are the same as those of Strategy C above: crops provide 62.9% of the diet, wild plants contribute 7.4%, and wild game fulfills the final 29.7%. The general pattern of resource exploitation is similar to that of Strategy C, but there are some differences. The proportion of red deer is lower during the winter, and the proportion of boar shows an increase throughout the fall and winter. Small game is less important in the late fall/early winter and throughout the spring, and summer. Fish, however, is more important in October, from March through May, and in July and August.

H. Ovicaprid Resources

This strategy corresponds to Strategy E above. Goat and sheep resources contribute 9% of the annual diet, crops provide 62.9%, wild plants add 5.6%, and wild game fulfills the remaining 22.5%. The effects of a lowered red deer density are apparent from the shape of the resource exploitation curves. Red deer is not only less important throughout the fall and winter, it no longer has the mid-winter bimodality apparent with the normal red deer density. Boar is slightly more important in February and

April. Small game is more important from October through January, and the importance of fish increases in May and June.

Comparative Efficiencies of Strategies E–H

Lowered densities result in the decreased exploitation of red deer, and in some cases the annual pattern of exploitation changes. The significance of changes in small game, boar, and fish exploitation vary from strategy to strategy depending on the proportion of domestic resources in the diet. As before, the amount and minimum territories were calculated for game (Tables 67 and 68) and wild plants (Tables 69 and 70). Comparing Tables 63 and 67, the annual total of red deer and boar that must be taken actually changes little with a decreased red deer density. However, this is not the case for fish and small game. In all cases the requisite amounts of fish increased. This has ramifications in determining the minimal territory needed for the village.

Decreasing the red deer density almost doubles the amount of territory needed for hunting red deer. But fish, not red deer, establish minimum territory sizes. The minimum territory under a normal red deer density varies from 37 to 110.6 square kilometers. Even at a density of two individuals per square kilometers, this would allow for an annual take of 15 to 21 red deer. If a change in red deer density only affected red deer, an increase in territory size would not be necessary. However, a decrease in red deer density effects an increase in the proportion of fish in the annual diet. If fish are the limiting factor in determining territory size, territories must be increased by 7% to allow for sufficient fish exploitation. The most logical alternative to increasing territory size, of course, is to completely restructure wild resource exploitation in order to decrease the significance of fish in the diet.

Table 67

Farmer Game Requirements: Low Red Deer Densities
(expressed as number of individuals)

Strategy	Red Deer	Roe Deer	Boar	Fish†	Beaver	Small Game
(E) Optimal	3.3	2.8	1.9	776.3	0.5	317.2
(F) No milk	6.7	5.7	3.0	1360.6	1.3	317.0
(G) No livestock	11.7	9.8	6.2	2322.4	2.3	384.1
(H) Ovicaprids	8.8	7.4	4.4	1751.6	1.7	317.3

†Fish are assumed to average one kilogram per individual

Table 68

Minimum Territory Needed for Game Resources:
Farmers, Low Red Deer Densities
(km^2)

Strategy	Red Deer	Roe Deer	Boar	Fish	Beaver	Small Game
(E) Optimal	8.2	1.2	0.8	39.6	3.7	15.4
(F) No milk	16.7	2.4	1.2	69.4	10.2	15.4
(G) No livestock	29.2	4.1	2.6	118.5	18.0	18.6
(H) Ovicaprids	22.9	3.1	1.8	89.4	13.3	15.4

Table 69

Farmer Annual Requirements of Wild Plant Foods: Low Red Deer Densities (kg)

Resource	Strategy			
	Optimal (E)	No Milk (F)	No Livestock (G)	Ovicaprids (H)
Tubers	32.4	95.6	205.8	122.1
Greens	60.2	124.9	240.5	159.2
Seeds	74.1	112.5	140.0	147.8
Berries	90.2	141.8	237.4	180.0
Hazel nuts	98.3	136.9	254.3	194.8

Crops and Livestock as a Forager Resource

This section examines the effects that differing levels of wheat in the annual diet would have on the foraging strategies of the hunter-gatherers. It seeks to determine the effect that the amount of wheat has on the minimal territory needed by the foragers, and to identify the points at which the foragers would be most likely to come into direct competition with the farmers.

The model developed in Chapter 6 suggests harvest yields varied annually, but sufficient wheat could be obtained regularly to provide the foragers with a 4-, 8-, or 12-week supply of wheat. These represent 3.8%, 7.7%, and 11.5% of the annual diet respectively (Table 71). These percentages were used to evaluate the effects of wheat on forager subsistence strategies, and how the addi-

Table 70

Territory Needed for Farmer Plant Exploitation
Low Red Deer Densities

Resource	Strategy			
	Optimal (E)	No Milk (F)	No Live-stock (G)	Ovi-caprids (H)
Tubers (km shoreline)	0.24	0.71	1.54	0.91
Greens (ha)	4.97	10.32	19.87	13.16
Seeds (ha)	0.19	0.29	0.36	0.38
Berries (km forest edge)	3.34	5.25	8.79	6.66
Hazel nuts (km forest edge)	2.17	3.00	5.63	4.31

tion of wheat would affect exploitation patterns. In all, 14 strategies were examined: 7 with normal red deer densities, and 7 with low red deer densities (Table 72).

The forager monthly resource fractions are listed in Table 73. The optimal farming strategy would have been able to provide meat to the foragers every March and April. Although the amount of meat the farmers would be able to provide totaled only 0.7% of the forager diet, it would have been consumed during a two-month period. The seasonal availability of this resource resulted in a resource use schedule that differed from the one obtained when domestic meats were not available.

Table 71

Forager Subsistence Strategies
(% of annual diet)

Strategy and Wheat Supply	Domestic		Wild	
	Crops	Live-stock	Plants	Game
Normal Red Deer Densities				
(I) No Domestic Resources	0.0	0.0	20.0	80.0
With livestock resources				
(J) 4-week	3.8	0.7	20.0	75.5
(K) 8-week	7.7	0.7	20.0	71.6
(L) 12-week	11.5	0.7	20.0	67.8
Without livestock resources				
(M) 4-week	3.8	0.0	20.0	75.5
(N) 8-week	7.7	0.0	20.0	71.6
(O) 12-week	11.5	0.0	20.0	67.8
Low Red Deer Densities				
(P) No domestic resources	0.0	0.0	20.0	80.0
Without livestock resources				
(Q) 4-week	3.8	0.7	20.0	76.2
(R) 8-week	7.7	0.7	20.0	72.3
(S) 12-week	11.5	0.7	20.0	68.5
Without domestic meat				
(T) 4-week	3.8	0.0	20.0	76.2
(U) 8-week	7.7	0.0	20.0	72.3
(V) 12-week	11.5	0.0	20.0	68.5

Multiplying the annual proportions of the various forager strategies (Table 72, Strategies I—O) by the monthly resource fractions (Table 73), monthly diets (Appendix: I-O) are obtained. Strategies I through M are based on normal red deer densities. Strategy I includes no domestic resources. Strategies J, K, and L are for diets respectively with a 4-, 8-, and 12-week supply of wheat. As discussed above, the optimal Neolithic strategy would have provided the foragers with livestock resources totalling 0.6%

Table 72

Forager Resource Proportions

I. Normal Red Deer Densities
(% of Annual Diet)

Resource	No Domestic Resources	With Livestock Resources			No Livestock Resources		
		Wheat Supply (no. of weeks)			Wheat Supply (no. of weeks)		
		4	8	12	4	8	12
	(I)	(J)	(K)	(L)	(M)	(N)	(O)
Red Deer	25.7	24.2	22.9	21.6	24.4	23.1	21.9
Roe Deer	3.0	2.9	2.8	2.6	3.0	2.8	2.6
Boar	22.1	20.8	19.7	18.6	21.0	19.9	18.8
Fish	13.1	12.3	11.7	11.1	12.5	11.8	11.2
Beaver	1.2	1.1	1.0	1.0	1.1	1.1	1.0
Small Game	12.9	12.2	11.5	10.9	12.2	11.6	11.0
Birds	2.0	2.0	2.0	2.0	2.0	2.0	2.0
Plants	20.0	20.0	20.0	20.0	20.0	20.0	20.0
Livestock	0.0	0.7	0.7	0.7	0.0	0.0	0.0
Wheat	0.0	3.8	7.7	11.5	3.8	7.7	11.5

of their annual diet. Strategies J through L include these livestock resources. Strategies M, N, and O are for the same diets, but do not.

Strategies P through V are calculated using low red deer densities; otherwise, they are comparable to Strategies I through O. Like Strategy I, Strategy P includes no domestic resources. Strategies Q, R, and S are for diets with 0.6% domestic meat and respectively a 4-, 8-, and 12-week supply of wheat. Finally, Strategies T, U, and V are calculated using the same levels of wheat and low red deer densities as Strategies Q, R, and S, but no domestic meat.

Table 72 (Continued)

II. Low Red Deer Densities
(% of Annual Diet)

Resource	No Domestic Resources	With Livestock Resources			Without Livestock Resources		
		Wheat Supply (in weeks)			Wheat Supply (in weeks)		
	(P)	4 (Q)	8 (R)	12 (S)	4 (T)	8 (U)	12 (V)
Red Deer	23.8	22.5	21.2	20.1	22.6	21.4	20.3
Roe Deer	3.1	2.9	2.8	2.6	3.0	2.8	2.7
Boar	22.6	21.3	20.2	19.1	21.5	20.4	19.3
Fish	14.0	13.2	12.5	11.8	13.4	12.7	12.0
Beaver	1.2	1.1	1.0	1.0	1.1	1.1	1.0
Small Game	13.3	12.5	11.9	11.2	12.6	11.9	11.2
Birds	2.0	2.0	2.0	2.0	2.0	2.0	2.0
Plants	20.0	20.0	20.0	20.0	20.0	20.0	20.0
Livestock	0.0	0.7	0.7	0.7	0.0	0.0	0.0
Wheat	0.0	3.8	7.7	11.5	3.8	7.7	11.5

Normal Red Deer Densities

Monthly subsistence requirements were determined as discussed above and are presented in Appendix I. Figure 10 shows the resource use schedule for Strategies I-O; the resource exploitation patterns for I-L and for M-O are presented in Figures 11 and 12.

I. No Domestic Resources

In this strategy wild plants provide 20% of the diet, fowl provide another 2% of the diet, and wild game contribute the remaining 78%. Boar are significant from late fall through midwinter, red deer are important throughout the winter, fish are sig-

Table 73

Forager Monthly Resource Fractions
(% of resource annual total)

Resource	Sept	Oct	Nov	Dec	Jan	Feb	March	April	May	June	July	Aug
Red Deer	8.5	8.0	6.5	7.5	15.0	14.0	15.0	6.5	4.0	3.5	5.5	6.0
Roe Deer	7.5	7.5	7.5	9.5	12.5	11.5	12.0	8.5	6.0	5.5	5.5	6.5
Boar	7.5	7.5	13.5	13.5	12.0	10.0	10.0	4.0	4.0	5.5	6.0	6.5
Fish	14.0	3.0	2.0	2.0	1.5	2.0	3.5	7.0	25.0	14.0	13.0	13.0
Beaver	0.0	0.0	0.0	20.0	20.0	20.0	20.0	20.0	0.0	0.0	0.0	0.0
Birds	0.0	25.0	25.0	0.0	0.0	0.0	25.0	25.0	0.0	0.0	0.0	0.0
Wild Plants	12.5	12.5	12.5	0.0	0.0	0.0	0.0	12.5	12.5	12.5	12.5	12.5
Livestock	0.0	0.0	0.0	0.0	0.0	0.0	60.0	40.0	0.0	0.0	0.0	0.0
Wheat	0.0	0.0	10.0	10.0	0.0	0.0	40.0	40.0	0.0	0.0	0.0	00

nificant in the late spring, and wild plant foods are important throughout the summer and fall. Small game are significant only in the early winter.

J. 4-Week Wheat Supply with Livestock Resources

The resource use schedule changes only slightly with the addition of a 4-week wheat supply. The amount of all resources decreases; nonetheless, their exploitation patterns remain the same. Boar continue to be important from late fall through early winter, and red deer are important throughout the winter. Plants become significant resources in March, and their importance does not diminish until the late fall. The proportion of small game increases in the fall, winter, and summer; but because of the wheat, there is a hiatus in small game exploitation in March. Finally, fish are less important in the winter, spring, and summer; nonetheless they continue to be a critical spring and summer resource.

K. 8-Week Wheat Supply with Livestock Resources

Domestic resources now provide 8.4% of the diet, and wild resources provide the remaining 91.6%. Boar continue to be an important late fall and early winter resource, although their importance decreases in mid-winter. By March boar are no longer exploited. Red deer remain an important winter resource, but the absolute percentage of red deer decreases. The winter exploitation of small game has also changed. In the previous two strategies, small game were important early winter resources, although their significance decreased throughout the winter. This is no longer so: their importance increases in the late winter, decreases sharply in the spring, and rises again in the summer. The importance of fish decreases slightly throughout the spring, but otherwise remains unchanged. Finally, plants do not become a primary resource until April, but thereafter they continue to be important throughout the summer and fall.

L. 12-Week Wheat Supply with Livestock Resources

In this strategy, wild resources provide 87.8% of the diet. Changes in the exploitation patterns of each game resource are readily apparent. Boar make their greatest contribution to the diet in December, but their significance decreases sharply thereafter, and they not exploited in March. Red deer continue to be an important winter resource; however, they are no longer an important March resource, and their proportion of the spring and

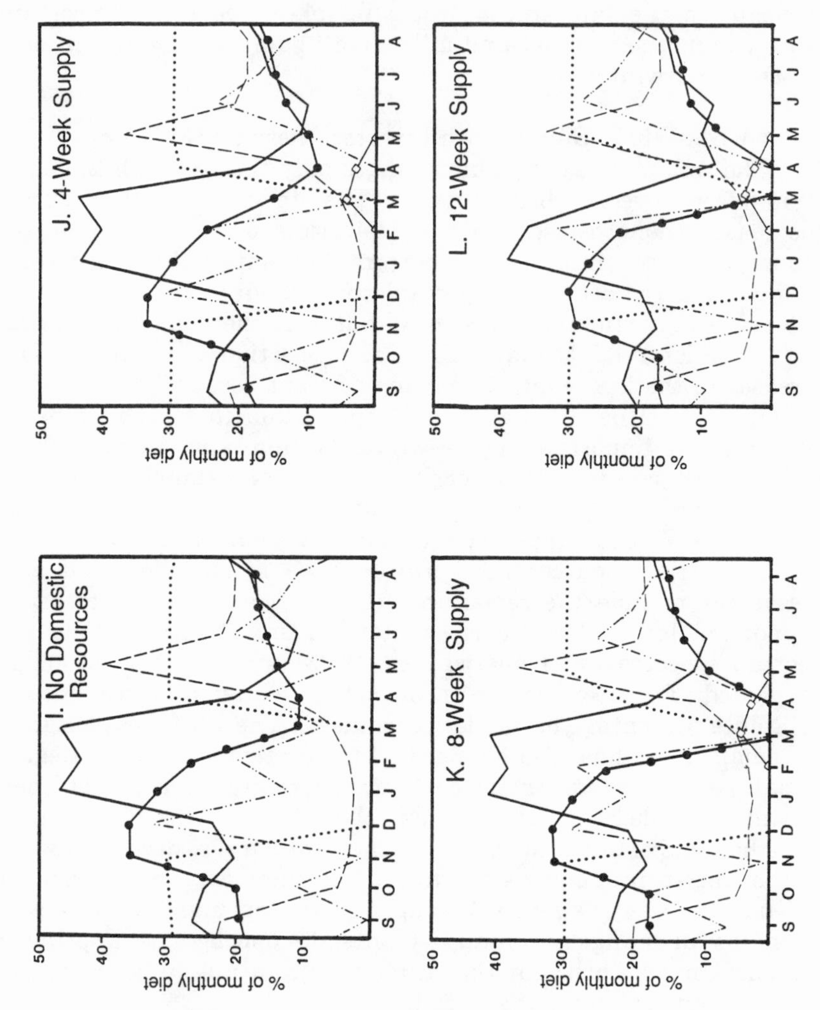

Figure 10. Forager Resource Use Schedule: Strategies I-O

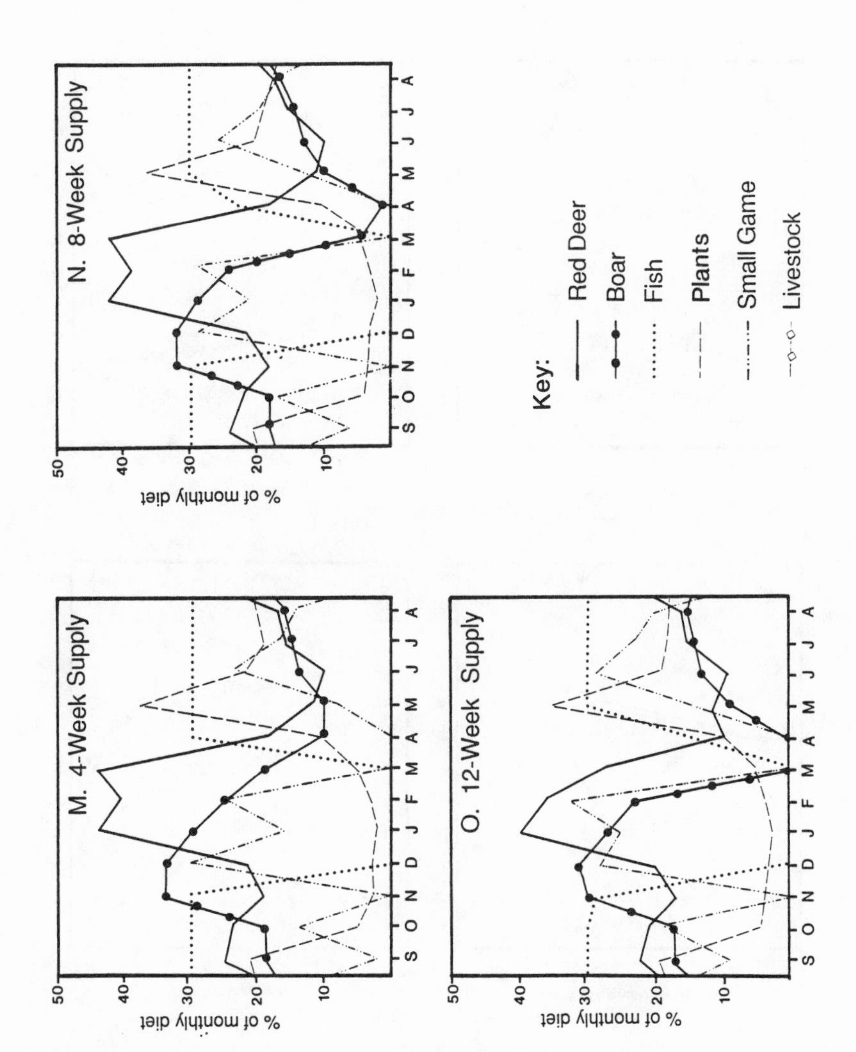

Figure 10 (continued).

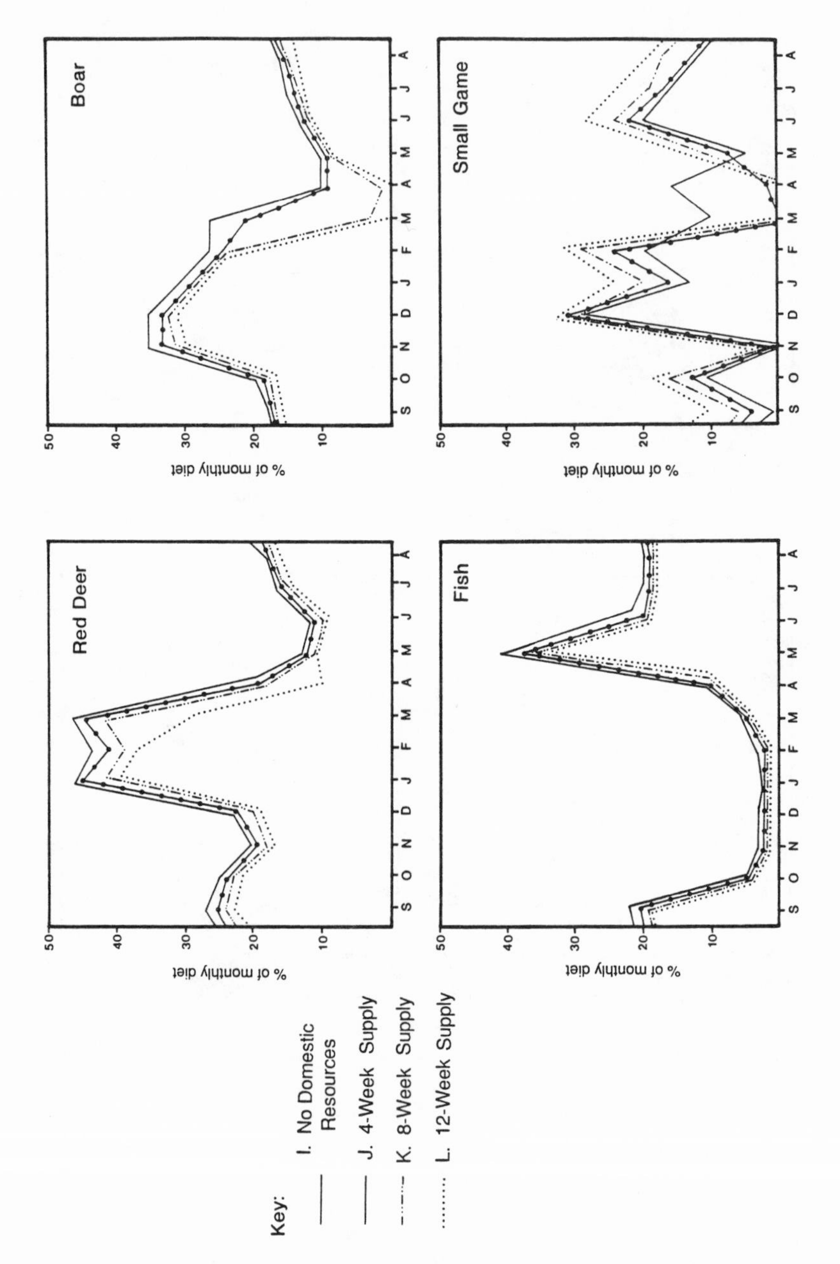

Figure 11. Forager Exploitation Patterns: Strategies I-L

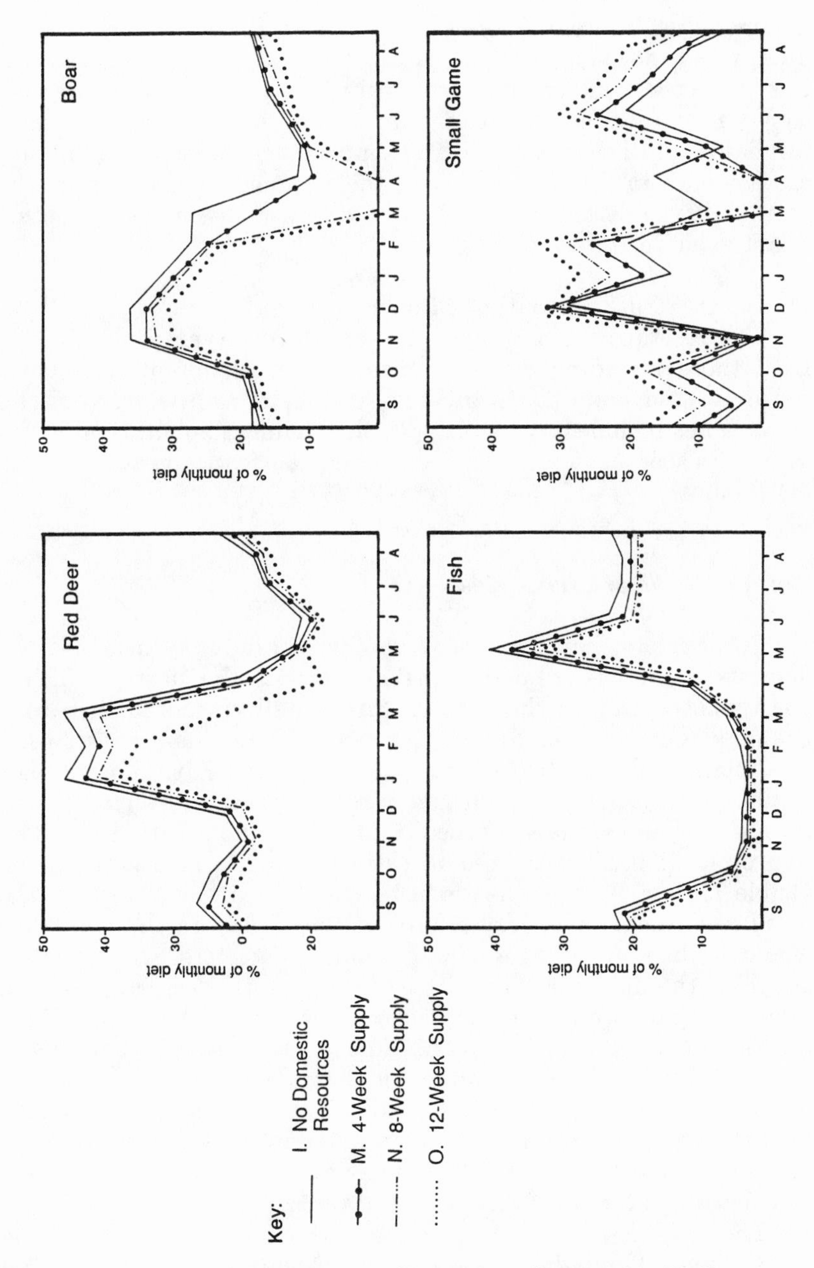

Figure 12. Forager Exploitation Patterns: Strategies M-O

summer diet is greatly reduced. The significance of fish is also diminished, although not to the same degree. Wild plants play a less important role in the early spring diet, but by May they become a significant resource, and their importance continues unabated until November. The importance of small game increases both in the winter and in the summer, but because of the availability of domestic resources in March and April, small game is not exploited in the spring.

M., N., O. Strategies Without Domestic Meat

The resource use patterns of each strategy vary only slightly from those including domestic meat. The consumption of boar, red deer, and plant foods increases in the spring when domestic meat is not included. In addition, the summer exploitation of fish also increases as a result of excluding domestic meat from the spring diet.

Comparisons of Strategies I–O

The exploitation of all wild resources gradually decreases as domestic resources are added to the diet, with the most significant changes occurring in the spring. First small game and then boar are no longer exploited during this season; later the proportions of wild plant foods and of red deer decrease markedly. The proportional contributions of each resource have been converted to the number of individuals needed annually (Table 74), and these figures have been converted to estimate territorial requirements (Table 75). Fish are the critical resource in determining the forager minimum territory size. With no wheat in the diet, they would require an area of 209.5 square kilometers. As wheat is added to the diet, the annual fish requirements decrease. With a 4-week wheat supply, territory size can be reduced by 6.9%. When an 8-week supply of wheat is obtained, territories can be reduced by 12.3%. If a 12-week supply is available, territories can be reduced by 15.5%. Thus as a result of interaction with the farmers, the foragers can reduce their territory size by as much as 15.5%. If domestic meat is not included in the forager diet, then the maximum territory reduction is 14.8%.

The use of wild plants by Mesolithic populations is poorly understood. Throughout this exercise it has been assumed that plants provide 20% of the annual diet, and that they are con-

Table 74

Forager Game Requirements: Normal Red Deer Densities
(expressed as number of individuals)

Strategy and Wheat Supply	Red Deer	Roe Deer	Boar	Fish†	Bea- ver	Small Game
(I) No Domestic Resources	24.1	18.7	11.9	4105.5	4.3	658.7
With Livestock Resources						
(J) 4-week	22.7	17.6	10.7	3873.9	4.1	667.8
(K) 8-week	21.5	16.7	9.1	3668.6	3.8	771.9
(L) 12-week	18.5	15.8	8.6	3468.6	3.6	876.8
Without Livestock Resources						
(M) 4-week	22.9	17.8	11.1	3905.5	4.0	646.3
(N) 8-week	21.7	16.8	9.4	3700.2	3.8	752.3
(O) 12-week	19.1	15.9	8.7	3500.2	3.6	857.1

†Fish are assumed to average one kilogram per individual

sumed from March through November. As discussed above, five classes of plant foods have been arbitrarily identified. The requisite amount of each plant resource has been estimated and is presented in Table 76, and the foragers territorial requirements are presented in Table 77. Sufficient wild plant foods would be available within the core area of the hunting territory. Nonetheless, the hazel requirement calls for nearly eleven kilometers of forest edge and the berries require slightly more than twelve. An approximately three kilometer stretch of shoreline is needed to obtain sufficient tubers.

While including domestic resources in the diet of the foragers can reduce the minimum territory size by up to 15.5%, cultivation and stockbreeding activities may have reduced red deer densities in the vicinity of a village. Since red deer provide up to 25.7% of the foragers' annual diet, a change in their density may have had

Table 75

Minimum Forager Territory: Normal Red Deer Densities
(km^2)

Strategy and Wheat Supply	Red Deer	Roe Deer	Boar	Fish	Beaver	Small Game
(I) No Domestic Resources	30.1	7.8	4.9	209.5	33.2	31.9
With Livestock Resources						
(J) 4-week	28.4	7.4	4.5	197.6	31.3	32.4
(K) 8-week	26.9	6.9	3.8	187.2	29.7	37.5
(L) 12-week	23.1	6.6	3.6	177.0	28.1	42.6
Without Livestock Resources						
(M) 4-week	28.6	7.4	4.6	199.3	31.6	31.4
(N) 8-week	27.1	7.0	3.9	188.8	29.9	36.5
(O) 12-week	23.8	6.6	3.6	178.6	28.4	41.6

an even more profound affect on the hunter-gatherers than it had on the farmers. The following discussion examines the nature and intensity of these effects.

The Effects of Low Red Deer Density on Forager Resource Exploitation

Strategy P corresponds to Strategy I; and Strategies Q—S correspond to J through L respectively, with the exception that calculations were made using lowered red deer densities. Similarly Strategies T—V correspond to Strategies K—M. The monthly diet of Strategies P—V are listed in the appendix. The resource use schedules with and without domestic meat are presented in Figure 13. The resource exploitation patterns are presented in Figures 14 and 15.

Table 76

Forager Annual Requirements of Wild Plant Foods
Normal Red Deer Densities
(kg)

Strategy and Wheat Supply	Tubers	Greens	Seeds	Berries	Hazel Nuts
(I) No Domestic Resources	432.9	234.9	405.5	330.9	491.4
With Livestock Resources					
(J) 4-week	426.98	232.60	406.21	330.90	494.70
(K) 8-week	386.60	216.60	405.70	330.90	492.50
(L) 12-week	347.14	200.90	404.80	330.90	488.00
Without Livestock Resources					
(M) 4-week	432.90	234.97	406.21	330.90	494.70
(N) 8-week	392.80	219.46	405.40	330.90	491.20
(O) 12-week	353.00	203.20	404.50	330.90	488.10

P. No Domestic Resources

As with Strategy I above, this is a pure hunting-gathering strategy with no reliance on domestic foodstuffs. The resource exploitation patterns and the resource use schedules are very similar to those of Strategy I. The primary difference between the two is that the proportion of red deer decreases slightly in the winter, while that of boar increases in the fall and early winter, and that of fish rises throughout spring and summer into into fall.

Q. 4-Week Wheat Supply and Livestock Resources

This corresponds to Strategy J above. Domestic resources provide 4.5% of the annual diet while wild plants, fowl, and game provide the remaining 95.5%. Boar and red deer continue to be

Table 77

Area Needed for Forager Plant Exploitation
Normal Red Deer Densities
(kg)

Strategy and Wheat Supply	Tubers†	Greens (ha)	Seeds (ha)	Berries‡	Hazel Nuts‡
(I) No domestic Resources	3.23	19.41	1.05	12.26	10.87
With Livestock Resources					
(J) 4-week	3.19	19.22	1.05	12.26	10.95
(K) 8-week	2.89	17.90	1.05	12.26	10.90
(L) 12-week	2.59	16.60	1.05	12.26	10.80
Without Livestock Resources					
(M) 4-week	3.23	19.42	1.05	12.26	10.95
(N) 8-week	2.93	18.14	1.05	12.26	10.87
(O) 12-week	2.64	16.79	1.05	12.26	10.80

†km shoreline ‡km forest edge

important resources in the fall and winter. However the significance of red deer decreases, whereas the importance of boar has increased over that of Strategy J. As with Strategy J, plant foods continue to be important throughout the spring, summer, and fall. The significance of small game differs from that of Strategy J: small game is slightly more important in the fall and early winter, but it is less important in the spring. The critical difference between the two strategies lies in the relative significance of fish. Fish provides a larger proportion of the diet throughout the whole year.

R. 8-Week Wheat Supply and Livestock Resources

This strategy includes a larger proportion of domestic resources in the annual diet. Trends that were identified previously are more marked here. The significance of red deer decreases in the winter, while the importance of both boar and small game increase over that of K. The significance of fish increases in all months.

S. 12-Week Wheat Supply and Livestock Resources

The effects of a low deer density become more evident as the annual proportion of wheat increases. In this case domestic resources fulfill 12.2% of the diet, wild resources contribute the final 87.8%. As is to be expected, the importance of red deer decreases and the significance of both boar and small game increases in the winter. But small game now provide a larger proportion of the late winter diet than red deer. Continuing this trend, the significance of fish increases slightly throughout the year.

T., U., V. Strategies With No Domestic Meat

As before, the exploitation of boar, red deer, and wild plant foods increases when domestic meats are not available in the spring. The amount of fish is once again affected in the late summer, and fish exploitation increases.

Comparisons of Strategies P–V

The game and territorial requirements are presented respectively in Table 78 and Table 79. The most significant effect of a low red deer density is found primarily among small game and fish. The exploitation pattern of small game varies from strategy to strategy, rather than exhibiting a gradual increase in importance in the diet. Fish, on the other hand, exhibit the same exploitation pattern, but their proportion of the monthly diet gradually increases. The proportion of wild plant foods in the annual diet remains the same regardless of the red deer densities. Consequently, there is no increase of plant foods in the diet, and the territorial requirements for plants remain as shown in Table 77. The most significant effects of a decreased red deer density, therefore, are a decreased reliance on red deer and a greater dependence on fish and small game. This increases the minimal territory size by 5.3% to 9.0% over that required with normal red deer densities. This increase is only partially offset by the addition of domestic resources to the forager diet.

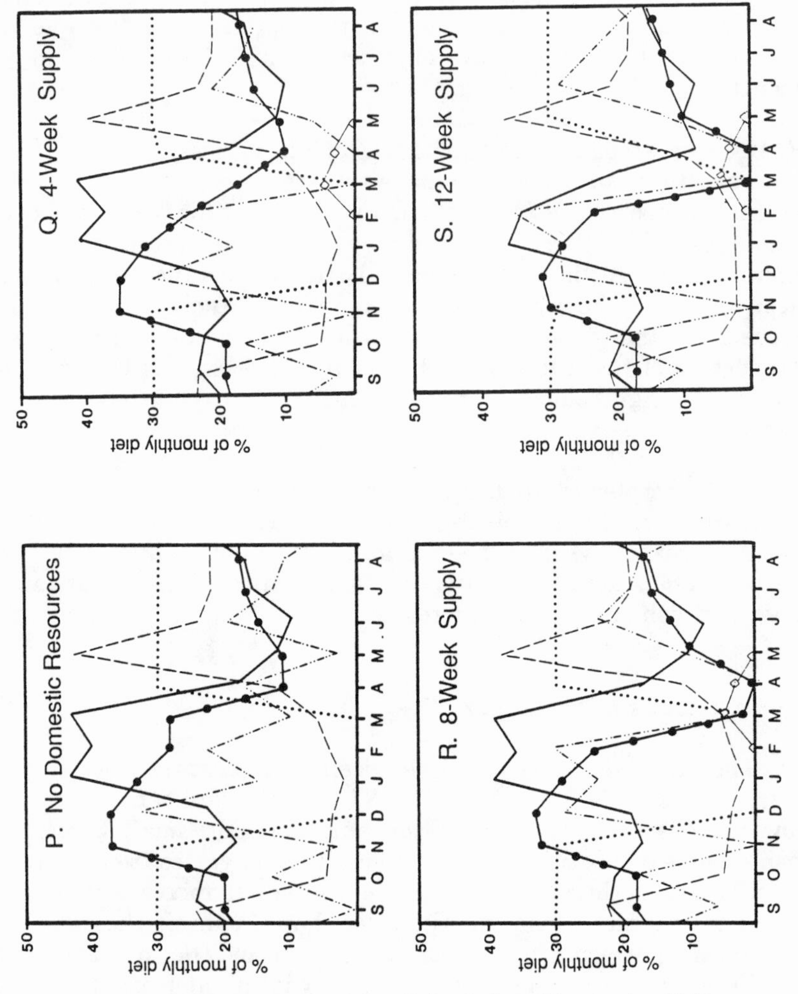

Figure 13. Forager Resource Use Schedule: Strategies P-V

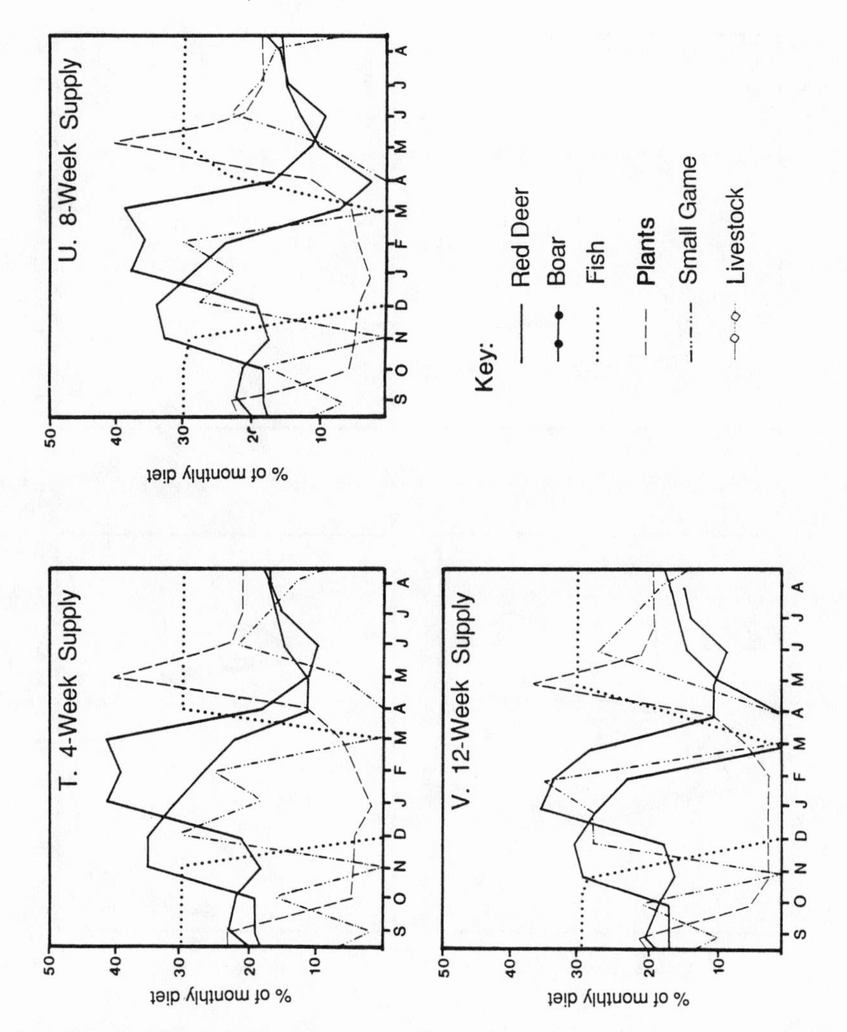

Figure 13 (continued).

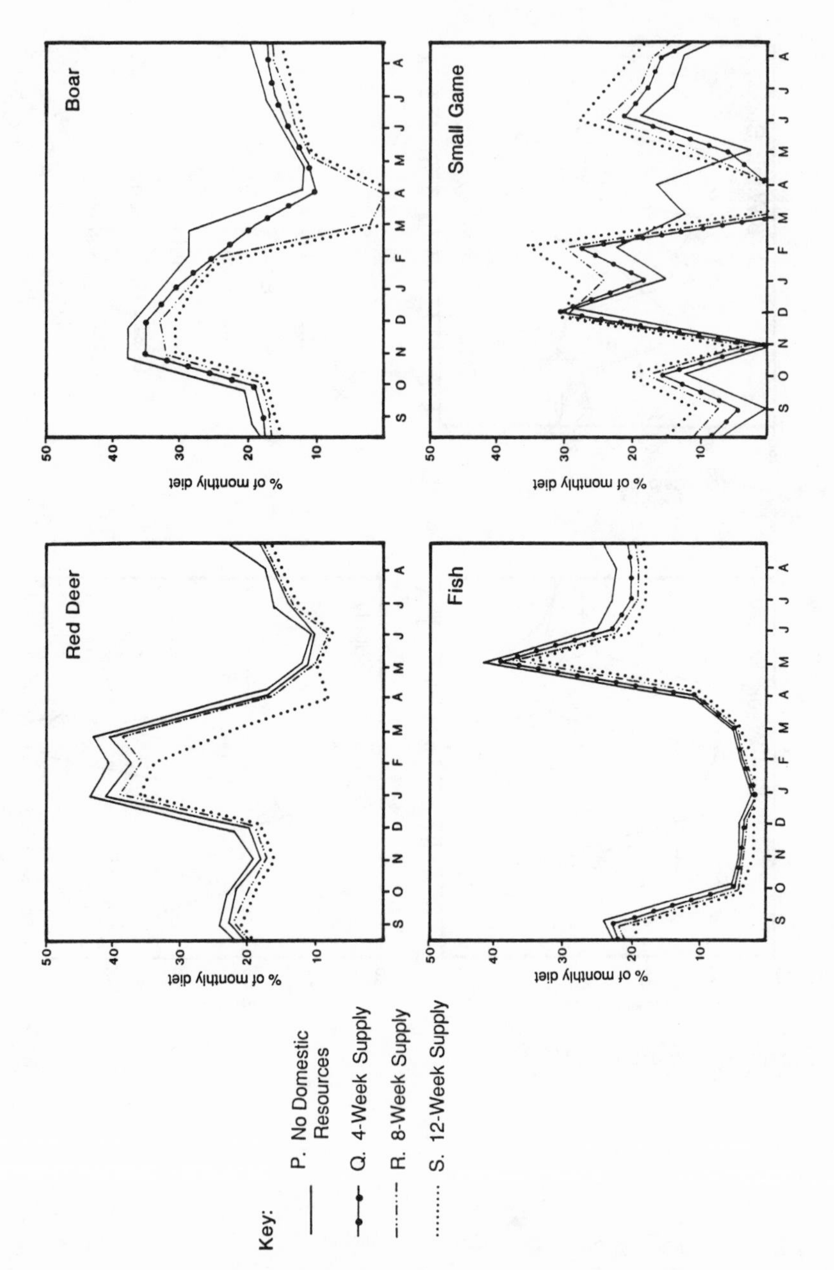

Figure 14. Forager Exploitation Patterns: Strategies P-S

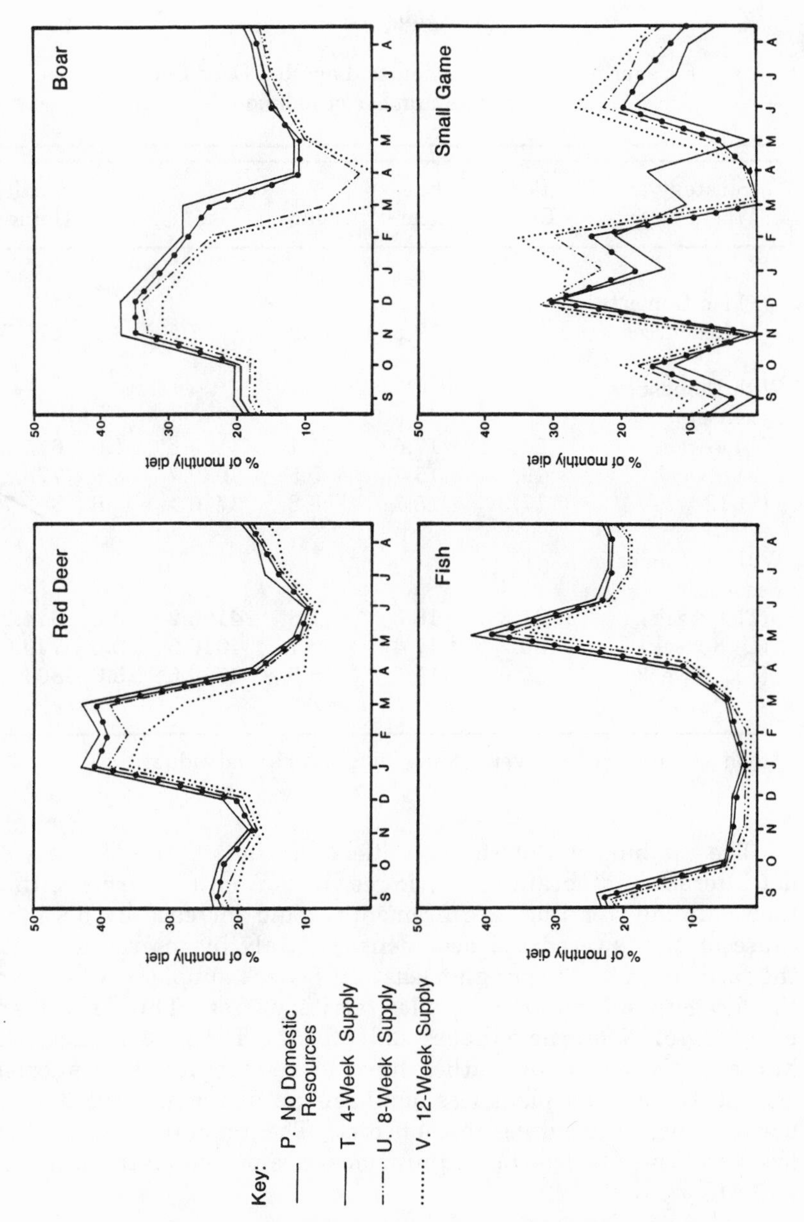

Figure 15. Forager Exploitation Patterns: Strategies P, T-V

Table 78

Forager Game Requirements: Low Red Deer Densities
(expressed as number of individuals)

Strategy and Wheat Supply	Red Deer	Roe Deer	Boar	Fish†	Bea-ver	Small Game
(P) No Domestic Resources	22.2	19.2	12.2	4417.5	3.6	675.7
With Domestic Resorces						
(Q) 4-week	21.1	17.6	11.1	4150.6	4.0	675.8
(R) 8-week	19.9	16.7	9.5	3930.7	3.8	778.2
(S) 12-week	17.0	15.8	8.8	3716.4	3.6	882.7
Without Domestic Resources						
(T) 4-week	21.3	18.6	11.5	4198.2	3.6	644.2
(U) 8-week	20.2	17.4	9.7	4010.3	3.6	749.0
(V) 12-week	17.6	15.6	8.9	3759.6	3.6	864.2

†Fish are assumed to average one kilogram per individual

The combination of the two effects is that, if the foragers do not interact and obtain domestic resources from the farmers, then their minimal territory requirements would increase by 6.9% because of the lowered red deer density. Only by interacting with the farmers and obtaining at least an 8-week supply of wheat are the foragers able to have smaller territory sizes. Thus, if the appearance of Neolithic villages did affect red deer densities, the hunter-gatherers would either have to restructure their subsistence strategies and place less emphasis on fish or incorporate the use of domestic resources in their diet. The amounts of wild plant foods and their territorial requirements are presented in Tables 80 and 81.

Table 79

Minimum Forager Territory: Low Red Deer Densities (km^2)

Strategy and Wheat Supply	Red Deer	Roe Deer	Boar	Fish	Beaver	Small Game
(P) No Domestic Resources	55.5	8.0	5.1	225.4	28.4	32.8
Without Livestock Resources						
(Q) 4-week	52.7	7.4	4.6	211.8	31.3	32.8
(R) 8-week	49.9	7.0	3.9	200.5	29.7	37.8
(S) 12-week	42.6	6.6	3.7	189.6	28.1	42.85
With Livestock Resources						
(T) 4-week	53.2	7.7	4.8	214.2	28.4	31.3
(U) 8-week	50.5	7.2	4.0	204.6	28.4	36.4
(V) 12-week	44.1	6.5	3.7	191.8	28.4	41.9

Forager-Farmer Interaction

The primary goal of this chapter has been to identify the benefits the foragers would have accrued from a mutualistic interaction with the farmers and to determine the extent to which behavioral modifications may have been needed for their participation in the relationship. Seasonality of domestic resource availability emerges as a critical factor in evaluating whether the foragers and farmers would develop a mutualistic or competitive interaction. Fish are the limiting factor in determining minimum territory sizes. Any increase or decrease in the amount of fish in the diet of either population results in a corresponding change in territory size. The significance of fish in either population's diet is affected by two factors. A decrease in the required amounts of fish results

Table 80

Forager Annual Requirements of Wild Plant Foods:
Low Red Deer Densities
(kg)

Strategy and Wheat Supply	Tubers	Greens	Seeds	Berries	Hazel Nuts
(P) No Domestic Resources	432.90	234.90	405.9	330.90	491.40
With Livestock Resources					
(Q) 4-week	427.84	232.96	406.21	330.90	494.68
(R) 8-week	387.51	216.95	405.92	330.90	493.35
(S) 12-week	348.23	201.34	404.98	330.90	489.03
Without Livestock Resources					
(T) 4-week	432.90	235.00	406.21	330.90	494.68
(U) 8-week	394.46	219.70	404.77	330.90	488.08
(V) 12-week	356.02	204.43	404.77	330.90	488.08

directly from the addition of domestic resources to the spring diet. A decrease also occurs indirectly, as a consequence of a general reduction in the absolute amount of wild game required in either population's diet. If domestic resources are added to the diet, reductions in the quantity of fish in the annual diet would occur regardless of when domestic resources were consumed. But when a superabundance of domestic resources is added during the same season in which fish are a critical resource, then the reduction is magnified.

This has several ramifications. First, the presence or absence as well as the quantity of domestic resources in the spring directly affects the proportion of fish in the farmer diet. Herd size and composition most directly control the availability of milk and meat. If, as discussed above, straw is used as winter fodder, then herd sizes and compositions will be influenced by the total number

Table 81

Area Needed for Forager Plant Exploitation: Low Red Deer Densities

Strategy and Wheat Supply	Tubers†	Greens (ha)	Seeds (ha)	Berries‡	Hazel Nuts‡
(P) No Domestic Resources	3.23	19.41	1.05	12.26	10.87
With Livestock Resources					
(Q) 4-week	3.20	19.25	1.05	12.26	10.95
(R) 8-week	2.90	17.93	1.05	12.26	10.92
(S) 12-week	2.60	16.64	1.05	12.26	10.82
Without Livestock Resources					
(T) 4-week	3.23	19.42	1.05	12.26	10.95
(U) 8-week	2.95	18.16	1.05	12.26	10.80
(V) 12-week	2.66	16.90	1.05	12.26	10.80

†km shoreline ‡km forest edge

of hectares planted in wheat. The best strategy for limiting competition between the farmers and foragers is for the farmers to plant enough wheat to ensure sufficient fodder for the optimal herds. But labor estimates suggest the farmers need the foragers' assistance in some years to plant their crops in a timely fashion. Cooperation on the part of the foragers is the only strategy that would reduce the territorial requirements of the farmers. At this level, cooperation is the best defensive strategy. Moreover, the addition of wheat and meat to the foragers' diet allows them to survive at higher population densities. Even if the farmers did not consume milk, domestic meats in the foragers' diet would reduce their exploitation of fish and game. Here again, cooperative interaction is the best defensive strategy.

Chapter 7

Summary and Conclusions

This chapter has considered how farming strategies might affect the hunter-gatherers through either indirect or direct competition. Direct competition would occur if the foragers incorporated large quantities of wild resources in their annual diets, while indirect competition would result from the farmers' changing their environment to such an extent that the resource base of the foragers would have been depleted. Four strategies were investigated: (1) the optimal strategy from Chapter 6; (2) use of meat but not milk; (3) no use of livestock; and (4) exploitation of pig and ovicaprid resources but not cattle. Each resulted in a different mix of domestic and wild resources in the diet. The structure of resource exploitation resulted from the seasonal availability of the domestic resources as well as the absolute quantity of each resource in the diet. In particular, the spring exploitation of fish proved to be the one point at which direct competition with the foraging population would develop.

A critical and unexpected result of this modeling showed that the optimal Neolithic subsistence strategy would provide a surplus of domestic meat in March and April. The availability of this meat is due to the death of calves, lambs, kids, ewes, does, and/or cows shortly before birth or within the first few weeks after birth. The surplus is thus a function of natural processes (not a result of slaughtering) and is to be expected every year.

The farmers would have been able to produce a regular and predictable surplus of wheat after 69% of the harvests, and they would have produced a surplus of meat and milk every spring. The production of these surpluses would have resulted from the farmers' regular activities, and it would have cost nothing to produce. The benefits derived from the hunter-gatherers would have ensured their continued existence as integral cultivators without having to develop a greater dependence on wild resources. For their parts, the hunter-gatherers would have received supplies of both grain and meat in the late winter or early spring, precisely when the quality of wild game would have been at a low point. Moreover, the addition of domestic resources would have allowed the foragers to live at a higher population density. In theory, the Mesolithic hunter-gatherers and the Neolithic cultivator-stockbreeders could have entered into a mutualistic interaction. Whether or not they actually *did* enter into such a relationship can only be determined from the archaeological record.

8
Archaeological Implications

A commonly held notion among archaeologists is that the foragers were incapable of competing with the farmers; consequently, their ultimate fate was to adopt farming or perish. Viewed diachronically over several centuries or a millennium, farmers were superior competitors; they evidently enjoyed a greater reproductive success and increased the size and density of their populations.

As Neolithic examples from northern Europe illustrate, tribally organized, residentially sedentary hunter-gatherer populations can effectively exclude farmers from their territories. Moreover, the ethnographic literature shows that stable and persistent, social relations can be maintained between residentially mobile hunter-gatherers and residentially sedentary cultivators. Interaction between two differently organized populations does not inevitably result in the amalgamation of the two. In fact, ethnic distinctions are often the very foundations on which large social systems are built (Barth 1969:11). This chapter examines how evidence of a long-term, stable relationship between the foragers and farmers might be examined using the archaeological record.

Testing the Model

The model is a static model of Mesolithic-Neolithic interaction; nonetheless diachronic implications can be drawn from it. First, cooperative interactions would be most likely to develop in an area occupied by mobile hunter-gatherers who have not yet begun to intensify their land use strategies. The appearance of farming populations would cause a minimal disruption in the forager's activities. At low population densities, interaction would not be significant to either population, but as population densities increased, competition would develop for territory to exploit wild resources. Competition between the two populations would arise not as a result of the farmers' degradation of the landscape, but as a direct consequence of their exploitation of wild game. As population densities increased, cooperative interaction would provide the foragers with a predictable source of domestic foodstuffs without interrupting their established subsistence round. As a direct consequence of this interaction, both populations would be able to reduce the size of their territories. Mutualistic interactions thus would allow both populations to survive at higher population densities.

The stability of this relationship is affected by two factors. The farmers benefit from the interaction because of the labor provided by the foragers. The foragers benefit from the reduced pressure on territory sizes resulting from a decreased reliance on fish resources by both populations. The stability of the relationship would be affected by (1) technological innovations—such as the introduction of either the plow or higher yielding wheats—resulting in decreased labor requirements for the farmers; (2) the farmers' consumption of domestic resources; or (3) environmental changes inhibiting the foragers' access to fish resources.

One of the most critical aspects of mutualistic interactions is that they can be formed and dissolved easily. As long as the farmers have no alternative use for the surplus wheat, it is to their advantage to trade it to the foragers. When alternative uses for wheat arise, it may become more desirable to use the wheat for the new purpose rather than trade it to the foragers. The worsening of the climate could also make it more advantageous to maintain larger emergency stores than to trade it to the foragers. Alternatively, the development of a hierarchial settlement system and the imposition of tribute would alter the value of the surplus wheat. Thus, the dissolution of a mutualistic relationship with the foragers could arise because of changing climatic conditions that made farming more unpredictable, or because of a changing social organization with a more desirable use (from the farmers' view) of the surplus wheat.

From a theoretical perspective, cooperative interaction should have existed between indigenous Mesolithic and immigrant Neolithic populations in the Alpine Foreland. But is there evidence to suggest that interaction did in fact occur? The answer to this question is a tentative yes. Since the problem has not been seriously considered before, there is a paucity of suitable data with which to determine the nature and intensity of the relationship. As discussed in Chapter 1, several problems exist in examining the Mesolithic/Neolithic transition. First, the Middle and Terminal Mesolithic are very poorly understood, and there is a lack of information on the social structure and economic organization of indigenous Mesolithic foragers prior to the appearance of Neolithic farmers. What evidence there is available suggests the Mesolithic populations of Southwest Germany had a high degree of residential mobility. The social organization and subsistence strategies of the earliest Neolithic populations in the region are only now beginning to be examined, although there is a long tradition of research at Middle Neolithic sites (Schmidt 1936).

At present the best evidence for interaction between Mesolithic and Neolithic populations comes from rock shelters and caves in the Swabian Alb. Most caves with mixed Mesolithic and Neolithic deposits, or with Neolithic components superimposed on Mesolithic occupations also show high frequencies of fish remains (Brunnacker *et al.* 1979; Taute 1966; Torke 1981). Few open air Mesolithic sites have been excavated, and the faunal assemblages at open air sites are poorly known. The most direct evidence of contact between the two populations is the occurrence of Mesolithic artifacts in Early Neolithic contexts. Refined excavation techniques are only now beginning to be used at some Early Neolithic villages; although it is impractical to screen all deposits, the recovery of microliths does occur.

Mounting circumstantial evidence suggests that contact and interaction occurred, but none is sufficient to demonstrate that it did in fact take place. Current data are now inadequate to test the model developed here, and future research must proceed in several directions to ensure suitable data are collected in the future. The most difficult aspect in examining cooperative farmer-forager interaction is that there may be little direct evidence of the interaction. If the model is correct, the foragers and farmers need only have been in contact with one another for a few weeks in the fall and early spring. To make matters more difficult, the items exchanged may have left few traces. If the foragers exchanged labor, then there would be no direct evidence in the archaeological record. Similarly, if the foragers received wheat from the farmers, evidence would survive in archeological contexts only if some of the grain was charred when cooked. Such occurrences would have been rare, and charred cereals cannot be expected in

Mesolithic contexts. Evidence of forager-farmer interaction must be sought simultaneously in several classes of data.

Settlement Pattern Analyses

If the model holds, then evidence of interaction should be visible in regional settlement patterns. The model suggests that foragers would be attracted to farming settlements in the spring and fall. The appearance of Neolithic villages in a region should act as a magnet, in effect attracting hunter-gatherer settlements at these two times of the year. Seasonality should be demonstrable through the faunal assemblage, particularly through the thin-sectioning of fish vertebrae. The most critical problem to resolve in such an analysis would be distinguishing between a temporary campsite of foragers and a campsite occupied by hunting farmers.

This problem could be resolved through two classes of data. First, the internal spatial organization of the two types of sites would be different. A forager site would accommodate family groups whereas a farmer site would in all likelihood accommodate adults—and probably only males at that. Surely both groups would use specialized tools. However, the hunting farmers most likely manufactured their tools in the village, while the foragers may have manufactured theirs at another temporary site, or at the source of raw material. In either event, a regional settlement study could be successful only by performing both a systematic survey and a series of excavations. Very few open air Mesolithic sites have been excavated, and there is little understanding of the Mesolithic settlement pattern. In order to determine whether Neolithic villages had a magnet effect, the regional settlement pattern of the Middle Mesolithic period would need to be established as a baseline for evaluating changes.

A mutualistic relationship could be demonstrated not only through a shifting of late winter/early spring sites toward Neolithic villages. As modeling of wild resource exploitation demonstrated, the entire annual resource use pattern would be affected by the addition of new resources in the spring, and Terminal Mesolithic sites should show a realignment of species exploitation, particularly fish and small game. If the model presented here is correct, then one would expect to see a decrease in the amount of small game exploited in the late winter months and an relative increase in the summer months. Thus, an understanding of changes that occur in Terminal Mesolithic settlement patterns can be established only with reference to Middle and Late Mesolithic settlement systems.

Regional settlement data also need to be collected for the Neolithic period. For the past 75 years, research has been highly biased towards the spectacular Middle Neolithic waterlogged villages and the cave sites. There is very little understanding of settlement patterns for other areas. Regional studies are needed to obtain information on Early Neolithic population density and to develop an understanding of the settlement pattern. Only then will it be possible to consider whether the foragers and farmers enjoyed a long-term stable interaction.

Exchange Networks

The model developed earlier is based on the exchange of a service in return for food. Both items are highly elastic. The farmers would not need forager labor every spring, and they would not be able to supply wheat every spring. Since a cooperative interaction can occur only if it is predictable, the relationship would have to be maintained through the exchange of inelastic resources. The exchange of marriage partners is the simplest and ethnographically the most common resource (Vierich 1982). Kinship entails reciprocal obligations, and through kinship both groups develop an obligatory responsibility to one another.

One aspect of a mutualistic interaction is that the majority of the members of the population participate. Here again ethnographic examples show an exchange of marriage partners as a common vehicle through which foragers and farmers maintained a formal relationship to each other. Such an exchange should be expected to have occurred during the Mesolithic/Neolithic transition if a long-term, mutualistic interaction developed. While it may be possible to identify marriage networks through the material remains, direct evidence of a marriage network would probably come from skeletal remains. The development of a common set of distinctive, genetic characteristics may allow a mixing of gene pools to be demonstrated. Moreover, it could be demonstrated through a common set of genetic characteristics in skeletal remains from each population.

The foragers may also have participated directly in exchange with the farmers. First, a number of resources would have been available in the forest that may not have been readily available to farmers living on its fringe. Evidence of such exchange would come primarily from faunal and floral remains. Honey may have been one such item. Honey would have been present—though very rare—in the deciduous forests of Southwest Germany. Although fruit and berries are available throughout the summer and can be dried for the winter, honey would have been a highly valued sweetener for the farmers. The foragers could have provided the farmers with honey. If it were stored in ceramic vessels rather than a leather pouch or birch bark container, then the contents could be identified through chemical and pollen analyses.

Finally, the foragers may have transported raw materials, finished products, and information from one region to another. Distribution of exotic materials would allow the existence of trade networks to be established. A comparative analysis of changes in Mesolithic and Neolithic periods would be needed to determine whether the Neolithic networks represented new patterns, or whether they were simply extensions of extant Mesolithic networks.

Paleopathological Studies

The most direct evidence of contact and interaction may come from the study of human skeletal populations. Hunter-gatherers often experience dietary stress in the late winter/early spring (Speth and Spielmann 1982). Evidence of such stress is often recorded in the human skeleton in the form of Harris lines and hypoplastic tooth enamel. If the foragers did indeed receive wheat from the farmers, then one would expect to see a decreased frequency of dietary stress in Terminal Mesolithic populations. If Mesolithic populations obtained milk from the farmers, then one might also expect to see an increase in the frequency of tuberculosis, a disease that affects bony tissues and is transmitted through milk. This is a wholly unexplored area.

Future Prospects

In the first few pages of this book it was suggested that traditional explanations of the Mesolithic/Neolithic transition overlooked the possibility that a long-term, stable relationship may have developed between indigenous foragers and immigrant farmers during the Neolithic colonization of Central Europe. In order to determine the posssibility of such a relationship, an interaction model was developed that suggested a mutualistic interaction could have been based on an exchange of services and domestic resources.

The model has been developed for use in Southwest Germany. It may be applicable to other areas of Europe; however, other regions of Europe experienced different social and environmental conditions during the Atlantic period. There are critical differences. First, the Mesolithic populations of Southwest Germany were residentially mobile hunter-gatherers. Unlike northern Europe (particularly Denmark and other coastal areas), where sedentary hunter-gathers had developed a relatively complex social organization, the Mesolithic foragers in Southwest Germany appear to have been an egalitarian, band level society.

Second, there were no anadromous fish runs in the area. This is in contrast to regions in which salmon runs provided a seasonal abundance of a resource. The lack of a superabundant resource in any given season means that the foragers would have had no storable resource. Instead, they would have been susceptible to seasonal lows in resource availability. This would have made them less able to compete directly with the farmers for territory; they would have been more vulnerable to indirect competition from the farmers, and they might have been more easily satisfied than would hunter-gatherer populations with a sedentary village lifestyle and abundant seasonal resources.

The methodological approach developed here should be applicable to examining forager-farmer interactions in other areas of Europe—even those areas in which direct and very intense competition is thought to have occurred. Keep in mind, however, that constraints affecting populations in other regions were probably significantly different. The difference would arise not only from differing structures of Mesolithic social organization and subsistence systems, but also from local constraints affecting Neolithic cultivation and stockbreeding. Future research should be directed towards highlighting the regional variability in Early Neolithic subsistence systems and identifying the contribution of Mesolithic populations to the successful adaptation of wheat farming to the deciduous forests of Central Europe.

While it is disappointing not to be able to resolve the problems of population interaction during the Mesolithic/Neolithic transition right now, the most critical areas of research are clear. Within the next decade many of the issues raised here should be resolved.

Appendix: Monthly Diets

Farmer Monthly Diet: Normal Red Deer Densities
(A) Optimal Strategy
(% of monthly diet)

Resource	Sept	Oct	Nov	Dec	Jan	Feb	Mar	Apr	May	Jun	Jul	Aug
Red Deer	3.9	3.6	2.9	3.4	6.8	6.3	0.0	0.0	1.2	1.6	2.4	2.6
Roe Deer	0.4	0.4	0.4	0.5	0.7	0.6	0.0	0.0	0.3	0.3	0.3	0.3
Boar	2.9	2.9	5.3	5.3	4.7	3.9	0.0	0.0	0.0	2.1	2.2	2.4
Fish	3.2	0.7	0.5	0.5	0.3	0.5	0.0	0.0	5.8	3.2	2.9	2.9
Beaver	0.0	0.0	0.0	0.4	0.4	0.4	0.0	0.0	0.0	0.0	0.0	0.0
Small Game	3.5	12.7	12.2	11.9	9.1	0.6	0.0	0.0	0.0	0.5	2.8	3.3
Birds	0.0	0.0	0.0	0.0	0.0	0.0	0.0	0.0	0.0	0.0	0.0	0.0
Plants	4.1	3.4	2.4	1.7	1.7	1.7	0.0	0.0	1.2	3.4	3.9	3.9
Livestock	10.6	13.4	13.4	13.4	13.4	23.1	25.8	24.9	13.6	11.0	10.1	10.1
Milk	8.5	0.0	0.0	0.0	0.0	0.0	14.3	14.3	15.0	15.0	12.5	11.6
Crops	62.9	62.9	62.9	62.9	62.9	62.9	62.9	62.9	62.9	62.9	62.9	62.9

Farmer Monthly Diet: Normal Red Deer Densities
(B) No Milk
(% of monthly diet)

Resource	Sept	Oct	Nov	Dec	Jan	Feb	Mar	Apr	May	Jun	Jul	Aug
Red Deer	5.9	5.6	4.5	5.2	10.5	9.8	8.6	4.3	2.8	2.4	3.7	4.0
Roe Deer	0.6	0.6	0.6	0.8	1.0	1.0	0.9	0.7	0.5	0.5	0.4	0.5
Boar	4.5	4.5	8.1	8.1	7.2	1.9	0.0	1.5	2.4	3.3	3.4	3.7
Fish	5.0	1.1	0.7	0.7	0.5	0.7	1.2	2.4	8.9	5.0	4.4	4.4
Beaver	0.0	0.0	0.0	0.6	0.6	0.6	0.6	0.6	0.0	0.0	0.0	0.0
Small Game	4.1	6.5	6.0	5.6	1.2	0.0	0.0	0.0	3.5	9.5	9.0	8.3
Birds	0.0	0.0	0.0	0.0	0.0	0.0	0.0	0.0	0.0	0.0	0.0	0.0
Plants	6.4	5.4	3.8	2.7	2.7	0.0	0.0	2.7	5.4	5.4	6.1	6.1
Livestock	10.6	13.4	13.4	13.4	13.4	23.1	25.8	24.9	13.6	11.0	10.1	10.1
Milk	0.0	0.0	0.0	0.0	0.0	0.0	0.0	0.0	0.0	0.0	0.0	0.0
Crops	62.9	62.9	62.9	62.9	62.9	62.9	62.9	62.9	62.9	62.9	62.9	62.9

Appendix

Farmer Monthly Diet: Normal Red Deer Densities
(C) No Livestock Products
(% of monthly diet)

Resource	Sept	Oct	Nov	Dec	Jan	Feb	Mar	Apr	May	Jun	Jul	Aug
Red Deer	10.1	9.5	7.7	8.9	17.9	16.7	17.0	7.4	4.8	4.2	6.2	6.8
Roe Deer	1.1	1.1	1.1	1.3	1.8	1.6	1.6	1.1	0.8	0.8	0.7	0.9
Boar	7.1	7.7	13.8	13.8	11.6	10.3	9.8	3.9	4.1	5.6	5.9	6.4
Fish	8.5	1.8	1.2	1.2	0.9	1.2	2.0	4.1	15.2	8.5	7.5	7.5
Beaver	0.0	0.0	0.0	1.1	1.1	1.1	1.0	1.0	0.0	0.0	0.0	0.0
Small Game	0.0	8.0	7.0	6.3	0.0	1.7	1.4	13.6	3.2	9.0	6.5	5.2
Birds	0.0	0.0	0.0	0.0	0.0	0.0	0.0	0.0	0.0	0.0	0.0	0.0
Plants	10.3	9.0	6.3	4.5	3.8	4.5	4.3	6.0	9.0	9.0	10.3	10.3
Livestock	0.0	0.0	0.0	0.0	0.0	0.0	0.0	0.0	0.0	0.0	0.0	0.0
Milk	0.0	0.0	0.0	0.0	0.0	0.0	0.0	0.0	0.0	0.0	0.0	0.0
Crops	62.9	62.9	62.9	62.9	62.9	62.9	62.9	62.9	62.9	62.9	62.9	62.9

Farmer Monthly Diet: Normal Red Deer Densities
(D) Ovicaprid Resorces
(% of monthly diet)

Resource	Sept	Oct	Nov	Dec	Jan	Feb	Mar	Apr	May	Jun	Jul	Aug
Red Deer	7.7	7.2	5.9	6.8	13.5	12.6	12.9	5.6	3.6	3.2	4.7	5.2
Roe Deer	0.8	0.8	0.8	1.0	1.3	1.2	1.2	0.9	0.6	0.6	0.6	0.7
Boar	5.8	5.8	10.5	10.5	9.3	7.8	0.7	3.0	1.0	4.3	4.4	4.8
Fish	6.5	1.4	0.9	0.9	0.7	0.9	1.5	3.1	11.5	6.5	5.7	5.7
Beaver	0.0	0.0	0.0	0.8	0.8	0.8	0.8	0.8	0.0	0.0	0.0	0.0
Small Game	0.9	12.4	11.5	11.0	5.4	7.0	0.0	1.9	0.0	0.0	3.2	6.1
Birds	0.0	0.0	0.0	0.0	0.0	0.0	0.0	0.0	0.0	0.0	0.0	0.0
Plants	8.2	6.8	4.8	3.4	3.4	3.4	0.0	4.6	4.7	6.8	7.8	7.8
Livestock	1.3	2.7	2.7	2.7	2.7	3.4	7.4	4.6	2.5	2.5	2.3	1.2
Milk	5.9	0.0	0.0	0.0	0.0	0.0	12.6	12.6	13.2	13.2	8.4	5.6
Crops	62.9	62.9	62.9	62.9	62.9	62.9	62.9	62.9	62.9	62.9	62.9	62.9

Farmer Monthly Diet: Low Red Deer Densities
(E) Optimal Diet
(% of monthly diet)

Resource	Sept	Oct	Nov	Dec	Jan	Feb	Mar	Apr	May	Jun	Jul	Aug
Red Deer	3.6	3.4	2.7	3.2	6.3	5.9	0.0	0.0	0.9	1.5	2.2	2.4
Roe Deer	0.4	0.4	0.4	0.5	0.7	0.6	0.0	0.0	0.3	0.3	0.3	0.3
Boar	3.0	3.0	5.4	5.4	4.8	4.0	0.0	0.0	0.0	2.2	2.3	2.5
Fish	3.5	0.7	0.5	0.5	0.4	0.5	0.0	0.0	6.2	3.5	3.1	3.1
Beaver	0.0	0.0	0.0	0.4	0.4	0.4	0.0	0.0	0.0	0.0	0.0	0.0
Small Game	3.4	12.8	12.3	12.0	9.4	0.9	0.0	0.0	0.2	0.2	2.7	3.2
Birds	0.0	0.0	0.0	0.0	0.0	0.0	0.0	0.0	0.0	0.0	0.0	0.0
Plants	4.1	3.4	2.4	1.7	1.7	1.7	0.0	0.0	0.9	3.4	3.9	3.9
Livestock	10.6	13.4	13.4	13.4	13.4	23.1	25.8	24.9	13.6	11.0	10.1	10.1
Milk	8.5	0.0	0.0	0.0	0.0	0.0	14.3	14.3	15.0	15.0	12.5	11.6
Cereal	62.9	62.9	62.9	62.9	62.9	62.9	62.9	62.9	62.9	62.9	62.9	62.9

Appendix

Farmer Monthly Diet: Low Red Deer Densities
(F) No Milk
(% of monthly diet)

Resource	Sept	Oct	Nov	Dec	Jan	Feb	Mar	Apr	May	Jun	Jul	Aug
Red Deer	5.5	5.2	4.2	4.9	9.7	9.1	8.5	4.0	2.6	2.3	3.4	3.7
Roe Deer	0.6	0.6	0.6	0.8	1.1	1.0	0.9	0.7	0.5	0.5	0.4	0.5
Boar	4.6	4.6	8.3	8.3	7.4	2.5	0.0	1.6	2.5	3.4	3.5	3.8
Fish	5.3	1.1	0.8	0.8	0.6	0.8	1.3	2.5	9.5	5.3	4.7	4.7
Beaver	0.0	0.0	0.0	0.6	0.6	0.6	0.6	0.6	0.0	0.0	0.0	0.0
Small Game	4.1	6.8	6.0	5.6	1.6	0.0	0.0	0.0	3.0	9.2	8.9	8.2
Birds	0.0	0.0	0.0	0.0	0.0	0.0	0.0	0.0	0.0	0.0	0.0	0.0
Plants	6.4	5.4	3.8	2.7	2.7	0.0	0.0	2.8	5.4	5.4	6.1	6.1
Livestock	10.6	13.4	13.4	13.4	13.4	23.1	25.8	24.9	13.6	11.0	10.1	10.1
Milk	0.0	0.0	0.0	0.0	0.0	0.0	0.0	0.0	0.0	0.0	0.0	0.0
Crops	62.9	62.9	62.9	62.9	62.9	62.9	62.9	62.9	62.9	62.9	62.9	62.9

Farmer Monthly Diet: Low Red Deer Densities
(G) No Livestock Resources
(% of monthly diet)

Resource	Sept	Oct	Nov	Dec	Jan	Feb	Mar	Apr	May	Jun	Jul	Aug
Red Deer	9.4	8.8	7.2	8.3	16.6	15.5	15.8	6.8	4.4	3.9	5.8	6.3
Roe Deer	1.1	1.1	1.1	1.4	1.8	1.7	1.7	1.2	0.9	0.8	0.8	0.9
Boar	7.3	7.9	14.2	14.2	12.4	10.5	10.0	4.0	4.2	5.8	6.0	6.5
Fish	9.1	2.0	1.3	1.3	1.0	1.3	2.2	4.4	16.3	9.1	8.1	8.1
Beaver	0.0	0.0	0.0	1.1	1.1	1.1	1.0	1.0	0.0	0.0	0.0	0.0
Small Game	0.0	8.3	7.0	6.3	0.0	2.5	2.1	13.7	2.3	8.5	6.1	5.0
Birds	0.0	0.0	0.0	0.0	0.0	0.0	0.0	0.0	0.0	0.0	0.0	0.0
Plants	10.2	9.0	6.3	4.5	4.2	4.5	4.3	6.0	9.0	9.0	10.3	10.3
Livestock	0.0	0.0	0.0	0.0	0.0	0.0	0.0	0.0	0.0	0.0	0.0	0.0
Milk	0.0	0.0	0.0	0.0	0.0	0.0	0.0	0.0	0.0	0.0	0.0	0.0
Crops	62.9	62.9	62.9	62.9	62.9	62.9	62.9	62.9	62.9	62.9	62.9	62.9

Farmer Monthly Diet: Low Red Deer Densities
(H) Ovicaprid Resources
(% of monthly diet)

Resource	Sept	Oct	Nov	Dec	Jan	Feb	Mar	Apr	May	Jun	Jul	Aug
Red Deer	7.1	6.7	5.4	6.2	12.5	11.7	11.9	5.2	3.3	2.9	4.4	4.8
Roe Deer	0.8	0.8	0.8	1.0	1.4	1.3	1.3	0.9	0.7	0.6	0.6	0.7
Boar	5.7	5.9	10.7	10.7	9.5	7.9	4.7	3.0	1.3	4.4	4.5	4.9
Fish	6.9	1.5	1.0	1.0	0.7	1.0	1.6	3.3	12.3	6.9	6.1	6.1
Beaver	0.0	0.0	0.0	0.8	0.8	0.8	0.8	0.8	0.0	0.0	0.0	0.0
Small Game	0.0	12.7	11.7	11.3	6.1	6.3	0.0	3.1	0.0	1.6	1.8	1.6
Birds	0.0	0.0	0.0	0.0	0.0	0.0	0.0	0.0	0.0	0.0	0.0	0.0
Plants	7.8	6.8	4.8	3.4	3.4	3.4	0.4	4.6	5.0	6.8	7.8	7.8
Livestock	2.2	2.7	2.7	2.7	2.7	4.7	5.2	5.0	2.8	2.2	2.1	2.1
Milk	6.6	0.0	0.0	0.0	0.0	0.0	11.2	11.2	11.7	11.7	9.8	9.1
Crops	62.9	62.9	62.9	62.9	62.9	62.9	62.9	62.9	62.9	62.9	62.9	62.9

Appendix

Forager Monthly Diet: Normal Red Deer Densities
(I) No Domestic Resources
(% of monthly diet)

Resource	Sept	Oct	Nov	Dec	Jan	Feb	Mar	Apr	May	Jun	Jul	Aug
Red Deer	26.3	24.8	20.1	23.2	46.4	43.3	46.4	20.1	12.4	10.8	17.0	18.6
Roe Deer	2.8	2.8	2.8	3.6	4.7	4.3	4.5	3.2	2.3	2.1	2.1	2.4
Boar	19.3	19.9	35.9	35.9	31.9	26.6	26.6	10.6	10.6	14.6	16.0	17.3
Fish	22.1	4.7	3.2	3.2	2.4	3.2	5.5	11.1	39.5	22.1	20.5	20.5
Beaver	0.0	0.0	0.0	2.8	2.8	2.8	2.8	2.8	0.0	0.0	0.0	0.0
Small Game	0.0	11.7	1.9	31.3	11.8	19.8	8.2	16.1	5.1	20.3	14.3	11.1
Birds	0.0	6.0	6.0	0.0	0.0	0.0	6.0	6.0	0.0	0.0	0.0	0.0
Plants	29.5	30.1	30.1	0.0	0.0	0.0	0.0	30.1	30.1	30.1	30.1	30.1
Livestock	0.0	0.0	0.0	0.0	0.0	0.0	0.0	0.0	0.0	0.0	0.0	0.0
Wheat	0.0	0.0	0.0	0.0	0.0	0.0	0.0	0.0	0.0	0.0	0.0	0.0

Forager Monthly Diet: Normal Red Deer Densities
(J) 4-Week Wheat Supply,
with Livestock Resources
(% of monthly diet)

Resource	Sept	Oct	Nov	Dec	Jan	Feb	Mar	Apr	May	Jun	Jul	Aug
Red Deer	24.8	23.3	19.0	21.7	43.8	40.8	43.8	19.0	11.7	10.2	15.9	17.5
Roe Deer	2.7	2.7	2.7	3.4	4.4	4.1	4.3	3.0	2.1	2.0	2.0	2.3
Boar	18.8	18.8	33.9	33.9	30.1	25.1	15.4	8.8	10.0	13.8	15.1	16.3
Fish	20.9	4.5	3.0	3.0	2.2	3.0	5.2	10.4	37.2	20.9	19.4	19.4
Beaver	0.0	0.0	0.0	2.7	2.7	2.7	2.7	2.7	0.0	0.0	0.0	0.0
Small Game	2.7	14.6	0.7	30.7	16.8	24.3	0.0	0.0	8.9	23.0	17.5	14.4
Birds	0.0	6.0	6.0	0.0	0.0	0.0	6.0	6.0	0.0	0.0	0.0	0.0
Plants	30.1	30.1	30.1	0.0	0.0	0.0	0.0	28.9	30.1	30.1	30.1	30.1
Livestock	0.0	0.0	0.0	0.0	0.0	0.0	4.3	2.9	0.0	0.0	0.0	0.0
Wheat	0.0	0.0	4.6	4.6	0.0	0.0	18.3	18.3	0.0	0.0	0.0	0.0

Forager Monthly Diet: Normal Red Deer Densities
(K) 8-Week Wheat Supply,
with Livestock Resources
(% of monthly diet)

Resource	Sept	Oct	Nov	Dec	Jan	Feb	Mar	Apr	May	Jun	Jul	Aug
Red Deer	23.5	22.1	18.0	20.7	41.4	38.7	41.1	18.0	11.1	9.7	15.2	16.6
Roe Deer	2.5	2.5	2.5	3.2	4.2	3.9	4.1	2.8	2.0	1.8	1.8	2.2
Boar	17.8	17.8	31.7	32.1	28.5	23.8	0.0	0.1	9.5	13.1	14.3	15.4
Fish	19.8	4.2	2.8	2.8	2.1	2.8	4.9	9.9	35.3	19.8	18.3	18.3
Beaver	0.0	0.0	0.0	2.5	2.5	2.5	2.5	2.5	0.0	0.0	0.0	0.0
Small Game	6.3	17.3	0.0	29.4	21.3	28.3	0.0	0.0	12.0	25.5	20.3	17.4
Birds	0.0	6.0	6.0	0.0	0.0	0.0	6.0	6.0	0.0	0.0	0.0	0.0
Plants	30.1	30.1	29.7	0.0	0.0	0.0	0.0	20.7	30.1	30.1	30.1	30.1
Livestock	0.0	0.0	0.0	0.0	0.0	0.0	4.3	2.9	0.0	0.0	0.0	0.0
Wheat	0.0	0.0	9.3	9.3	0.0	0.0	37.1	37.1	0.0	0.0	0.0	0.0

Forager Monthly Diet: Normal Red Deer Densities
(L) 12-Week Wheat Supply,
with Livestock Resources
(% of monthly diet)

Resource	Sept	Oct	Nov	Dec	Jan	Feb	Mar	Apr	May	Jun	Jul	Aug
Red Deer	22.2	20.9	16.9	19.6	39.2	36.5	23.4	8.6	10.5	9.1	14.4	15.7
Roe Deer	2.4	2.4	2.4	3.0	4.0	3.7	3.8	2.7	1.9	1.7	1.7	2.1
Boar	16.9	16.9	29.2	30.3	27.0	22.5	0.0	0.0	9.0	12.4	13.5	14.6
Fish	18.7	4.0	2.7	2.7	2.0	2.7	4.7	9.3	33.3	18.7	17.3	17.3
Beaver	0.0	0.0	0.0	2.4	2.4	2.4	2.4	2.4	0.0	0.0	0.0	0.0
Small Game	9.7	19.7	0.0	28.1	25.4	32.2	0.0	0.0	15.2	28.0	23.0	20.2
Birds	0.0	6.0	6.0	0.0	0.0	0.0	6.0	6.0	0.0	0.0	0.0	0.0
Plants	30.1	30.1	28.9	0.0	0.0	0.0	0.0	12.7	30.1	30.1	30.1	30.1
Livestock	0.0	0.0	0.0	0.0	0.0	0.0	4.3	2.9	0.0	0.0	0.0	0.0
Wheat	0.0	0.0	13.9	13.9	0.0	0.0	55.4	55.4	0.0	0.0	0.0	0.0

Forager Monthly Diet: Normal Red Deer Densities
(M) 4-Week Wheat Supply,
Without Livestock Resources
(% of monthly diet)

Resource	Sept	Oct	Nov	Dec	Jan	Feb	Mar	Apr	May	Jun	Jul	Aug
Red Deer	25.0	23.5	19.1	22.1	44.1	41.2	44.1	19.1	11.8	10.3	16.2	17.7
Roe Deer	2.7	2.7	2.7	3.4	4.5	4.1	4.3	3.0	2.1	2.0	2.0	2.3
Boar	19.0	19.0	34.2	34.2	30.4	25.3	19.3	10.1	10.1	13.9	15.2	16.4
Fish	21.0	4.5	3.0	3.0	2.3	3.0	5.3	10.5	37.5	21.0	19.5	19.5
Beaver	0.0	0.0	0.0	2.7	2.7	2.7	2.7	2.7	0.0	0.0	0.0	0.0
Small Game	2.2	14.2	0.3	30.0	16.0	23.7	0.0	0.2	8.4	22.7	17.0	14.0
Birds	0.0	6.0	6.0	0.0	0.0	0.0	6.0	6.0	0.0	0.0	0.0	0.0
Plants	30.1	30.1	30.1	0.0	0.0	0.0	0.0	30.1	30.1	30.1	30.1	30.1
Livestock	0.0	0.0	0.0	0.0	0.0	0.0	0.0	0.0	0.0	0.0	0.0	0.0
Wheat	0.0	0.0	4.6	4.6	0.0	0.0	18.3	18.3	0.0	0.0	0.0	0.0

Forager Monthly Diet: Normal Red Deer Densities
(N) 8-Week Wheat Supply,
Without Livestock Resources
(% of monthly diet)

Resource	Sept	Oct	Nov	Dec	Jan	Feb	Mar	Apr	May	Jun	Jul	Aug
Red Deer	23.7	22.3	18.2	20.9	41.8	39.0	41.8	18.1	11.1	9.8	15.3	16.7
Roe Deer	2.5	2.5	2.5	3.2	4.3	3.9	4.1	2.9	2.0	1.9	1.9	2.2
Boar	18.0	18.0	31.7	32.4	28.8	24.0	3.5	1.4	9.6	13.2	14.4	15.6
Fish	19.9	4.3	2.8	2.8	2.1	2.8	5.0	10.0	35.6	19.9	18.5	18.5
Beaver	0.0	0.0	0.0	2.5	2.5	2.5	2.5	2.5	0.0	0.0	0.0	0.0
Small Game	5.8	16.8	0.0	28.9	20.5	27.8	0.0	0.0	11.6	25.1	19.8	16.9
Birds	0.0	6.0	6.0	0.0	0.0	0.0	6.0	6.0	0.0	0.0	0.0	0.0
Plants	30.1	30.1	29.5	0.0	0.0	0.0	0.0	22.0	30.1	30.1	30.1	30.1
Livestock	0.0	0.0	0.0	0.0	0.0	0.0	0.0	0.0	0.0	0.0	0.0	0.0
Wheat	0.0	0.0	9.3	9.3	0.0	0.0	37.1	37.1	0.0	0.0	0.0	0.0

Forager Monthly Diet: Normal Red Deer Densities
(O) 12-Week Wheat Supply,
No Domestic Resources
(% of monthly diet)

Resource	Sept	Oct	Nov	Dec	Jan	Feb	Mar	Apr	May	Jun	Jul	Aug
Red Deer	22.4	21.2	17.1	19.8	39.5	36.9	27.6	10.0	10.5	9.2	14.5	15.8
Roe Deer	2.4	2.4	2.4	3.0	4.0	3.7	3.9	2.7	1.9	1.8	1.8	2.1
Boar	17.1	17.0	29.2	30.6	27.2	22.7	0.0	0.0	9.1	12.5	13.6	14.7
Fish	18.8	4.0	2.7	2.7	2.0	2.7	4.7	9.4	33.7	18.8	17.5	17.5
Beaver	0.0	0.0	0.0	2.4	2.4	2.4	2.4	2.4	0.0	0.0	0.0	0.0
Small Game	9.2	19.3	0.0	27.6	24.9	31.6	0.0	0.0	14.7	27.6	22.5	19.8
Birds	0.0	6.0	6.0	0.0	0.0	0.0	6.0	6.0	0.0	0.0	0.0	0.0
Plants	30.1	30.1	28.7	0.0	0.0	0.0	0.0	14.1	30.1	30.1	30.1	30.1
Livestock	0.0	0.0	0.0	0.0	0.0	0.0	0.0	0.0	0.0	0.0	0.0	0.0
Wheat	0.0	0.0	13.9	13.9	0.0	0.0	55.4	55.4	0.0	0.0	0.0	0.0

Forager Monthly Diet: Low Red Deer Densities
(P) No Domestic Resources
(% of monthly diet)

Resource	Sept	Oct	Nov	Dec	Jan	Feb	Mar	Apr	May	Jun	Jul	Aug
Red Deer	24.1	22.9	18.1	21.7	43.4	39.8	43.4	18.1	12.0	9.6	15.7	16.9
Roe Deer	2.4	2.4	2.4	3.6	4.8	4.8	4.8	3.6	2.4	2.4	2.4	2.4
Boar	19.9	20.5	37.3	37.3	32.5	27.7	27.7	10.8	10.8	14.5	16.9	18.1
Fish	24.1	4.8	3.6	3.6	2.4	3.6	6.0	12.0	42.2	24.1	21.7	21.7
Beaver	0.0	0.0	0.0	2.4	2.4	2.4	2.4	2.4	0.0	0.0	0.0	0.0
Small Game	0.0	13.3	2.5	31.4	14.5	21.7	9.7	17.0	2.5	19.3	13.2	10.8
Birds	0.0	6.0	6.0	0.0	0.0	0.0	6.0	6.0	0.0	0.0	0.0	0.0
Plants	29.5	30.1	30.1	0.0	0.0	0.0	0.0	30.1	30.1	30.1	30.1	30.1
Livestock	0.0	0.0	0.0	0.0	0.0	0.0	0.0	0.0	0.0	0.0	0.0	0.0
Wheat	0.0	0.0	0.0	0.0	0.0	0.0	0.0	0.0	0.0	0.0	0.0	0.0

Forager Monthly Diet: Low Red Deer Densities
(Q) 4-Week Wheat Supply,
With Livestock Resources
(% of monthly diet)

Resource	Sept	Oct	Nov	Dec	Jan	Feb	Mar	Apr	May	Jun	Jul	Aug
Red Deer	22.9	21.7	18.1	20.5	41.0	37.3	41.0	18.1	10.8	9.6	14.5	15.7
Roe Deer	2.4	2.4	2.4	3.6	4.8	3.6	4.8	2.4	2.4	2.4	2.4	2.4
Boar	19.3	19.3	34.9	34.9	31.3	25.3	16.9	9.6	10.8	14.5	15.7	16.9
Fish	22.9	4.8	3.6	3.6	2.4	3.6	6.0	10.8	39.8	22.9	20.5	20.5
Beaver	0.0	0.0	0.0	2.4	2.4	2.4	2.4	2.4	0.0	0.0	0.0	0.0
Small Game	2.4	15.7	0.1	30.2	18.1	27.8	0.0	0.1	6.1	20.5	16.8	14.4
Birds	0.0	6.0	6.0	0.0	0.0	0.0	6.0	6.0	0.0	0.0	0.0	0.0
Plants	30.1	30.1	30.1	0.0	0.0	0.0	0.0	28.9	30.1	30.1	30.1	30.1
Livestock	0.0	0.0	0.0	0.0	0.0	0.0	4.8	3.6	0.0	0.0	0.0	0.0
Wheat	0.0	0.0	4.8	4.8	0.0	0.0	18.1	18.1	0.0	0.0	0.0	0.0

Appendix

Forager Monthly Diet: Low Red Deer Densities
(R) 8-Week Wheat Supply,
with Livestock Resources
(% of monthly diet)

Resource	Sept	Oct	Nov	Dec	Jan	Feb	Mar	Apr	May	Jun	Jul	Aug
Red Deer	21.7	20.5	16.9	19.3	38.6	36.1	38.6	16.9	9.6	8.4	14.5	15.7
Roe Deer	2.4	2.4	2.4	3.6	3.6	3.6	3.6	2.4	2.4	2.4	2.4	2.4
Boar	18.1	18.1	31.9	32.5	28.9	24.1	2.4	0.0	9.6	13.3	14.5	15.7
Fish	21.7	4.8	3.6	3.6	2.4	3.6	4.8	10.8	37.3	21.7	19.3	19.3
Beaver	0.0	0.0	0.0	2.4	2.4	2.4	2.4	2.4	0.0	0.0	0.0	0.0
Small Game	6.0	18.1	0.1	29.0	24.1	30.2	0.1	0.1	11.0	24.1	19.2	16.8
Birds	0.0	6.0	6.0	0.0	0.0	0.0	6.0	6.0	0.0	0.0	0.0	0.0
Plants	30.1	30.1	29.5	0.0	0.0	0.0	0.0	20.5	30.1	30.1	30.1	30.1
Livestock	0.0	0.0	0.0	0.0	0.0	0.0	4.8	3.6	0.0	0.0	0.0	0.0
Wheat	0.0	0.0	9.6	9.6	0.0	0.0	37.3	37.3	0.0	0.0	0.0	0.0

Forager Monthly Diet: Low Red Deer Densities
(S) 12-Week Wheat Supply,
with Livestock Resources
(% of monthly diet)

Resource	Sept	Oct	Nov	Dec	Jan	Feb	Mar	Apr	May	Jun	Jul	Aug
Red Deer	20.5	19.3	15.7	18.1	36.1	33.7	22.9	7.8	9.6	8.4	13.3	14.5
Roe Deer	2.4	2.4	2.4	3.6	3.6	3.6	3.6	2.4	2.4	1.2	1.2	2.4
Boar	16.9	16.9	30.1	31.3	27.7	22.9	0.0	0.0	9.6	12.0	13.3	14.5
Fish	20.5	4.8	2.4	2.4	2.4	2.4	4.8	9.6	36.1	20.5	18.1	18.1
Beaver	0.0	0.0	0.0	2.4	2.4	2.4	2.4	2.4	0.0	0.0	0.0	0.0
Small Game	9.6	20.5	0.0	27.7	27.8	35.0	0.1	0.1	12.2	27.8	24.0	20.4
Birds	0.0	6.0	6.0	0.0	0.0	0.0	6.0	6.0	0.0	0.0	0.0	0.0
Plants	30.1	30.1	28.9	0.0	0.0	0.0	0.0	12.7	30.1	30.1	30.1	30.1
Livestock	0.0	0.0	0.0	0.0	0.0	0.0	4.8	3.6	0.0	0.0	0.0	0.0
Wheat	0.0	0.0	14.5	14.5	0.0	0.0	55.4	55.4	0.0	0.0	0.0	0.0

Forager Monthly Diet: Low Red Deer Densities
(T) 4-Week Wheat Supply,
Without Livestock Resources
(% of monthly diet)

Resource	Sept	Oct	Nov	Dec	Jan	Feb	Mar	Apr	May	Jun	Jul	Aug
Red Deer	22.9	21.7	18.1	20.5	41.0	38.6	41.0	18.1	10.8	9.6	14.5	16.9
Roe Deer	2.4	2.4	2.4	3.6	4.8	3.6	4.8	3.6	2.4	2.4	2.4	2.4
Boar	19.3	19.3	34.9	34.9	31.3	26.5	21.7	10.8	10.8	14.5	15.7	16.9
Fish	22.9	4.8	3.6	3.6	2.4	3.6	6.0	10.8	39.8	22.9	20.5	20.5
Beaver	0.0	0.0	0.0	2.4	2.4	2.4	2.4	2.4	0.0	0.0	0.0	0.0
Small Game	2.4	15.7	0.1	30.2	18.1	25.3	0.0	0.1	6.1	20.5	16.8	13.2
Birds	0.0	6.0	6.0	0.0	0.0	0.0	6.0	6.0	0.0	0.0	0.0	0.0
Plants	30.1	30.1	30.1	0.0	0.0	0.0	0.0	30.1	30.1	30.1	30.1	30.1
Livestock	0.0	0.0	0.0	0.0	0.0	0.0	0.0	0.0	0.0	0.0	0.0	0.0
Wheat	0.0	0.0	4.8	4.8	0.0	0.0	18.1	18.1	0.0	0.0	0.0	0.0

Forager Monthly Diet: Low Red Deer Densities
(U) 8-Week Wheat Supply,
Without Livestock Resources
(% of monthly diet)

Resource	Sept	Oct	Nov	Dec	Jan	Feb	Mar	Apr	May	Jun	Jul	Aug
Red Deer	21.7	20.5	16.9	19.3	38.6	36.1	38.6	16.9	10.8	9.6	14.5	15.7
Roe Deer	2.4	2.4	2.4	3.6	4.8	3.6	3.6	2.4	2.4	2.4	2.4	2.4
Boar	18.1	18.1	32.5	33.7	28.9	24.1	7.2	1.8	9.6	13.3	14.5	15.7
Fish	21.7	4.8	3.6	3.6	2.4	3.6	4.8	10.8	38.6	21.7	19.3	19.3
Beaver	0.0	0.0	0.0	2.4	2.4	2.4	2.4	2.4	0.0	0.0	0.0	0.0
Small Game	6.0	18.1	0.1	27.8	22.9	30.2	0.1	0.1	8.5	22.9	19.2	16.8
Birds	0.0	6.0	6.0	0.0	0.0	0.0	6.0	6.0	0.0	0.0	0.0	0.0
Plants	30.1	30.1	28.9	0.0	0.0	0.0	0.0	22.3	30.1	30.1	30.1	30.1
Livestock	0.0	0.0	0.0	0.0	0.0	0.0	0.0	0.0	0.0	0.0	0.0	0.0
Wheat	0.0	0.0	9.6	9.6	0.0	0.0	37.3	37.3	0.0	0.0	0.0	0.0

Forager Monthly Diet: Low Red Deer Densities
(V) 12-Week Wheat Supply,
Without Livestock Resources
(% of monthly diet)

Resource	Sept	Oct	Nov	Dec	Jan	Feb	Mar	Apr	May	Jun	Jul	Aug
Red Deer	20.5	19.3	15.7	18.1	36.1	33.7	27.7	9.6	9.6	8.4	13.3	14.5
Roe Deer	2.4	2.4	2.4	3.6	3.6	3.6	3.6	2.4	2.4	1.2	1.2	2.4
Boar	16.9	16.9	30.1	31.3	27.7	22.9	0.0	0.0	9.6	13.3	14.5	15.7
Fish	20.5	4.8	2.4	2.4	2.4	2.4	4.8	9.6	36.1	20.5	19.3	19.3
Beaver	0.0	0.0	0.0	2.4	2.4	2.4	2.4	2.4	0.0	0.0	0.0	0.0
Small Game	9.6	20.5	0.0	27.7	27.8	35.0	0.1	0.1	12.1	26.5	21.6	18.0
Birds	0.0	6.0	6.0	0.0	0.0	0.0	6.0	6.0	0.0	0.0	0.0	0.0
Plants	30.1	30.1	28.9	0.0	0.0	0.0	0.0	14.5	30.1	30.1	30.1	30.1
Livestock	0.0	0.0	0.0	0.0	0.0	0.0	0.0	0.0	0.0	0.0	0.0	0.0
Wheat	0.0	0.0	14.5	14.5	0.0	0.0	55.4	55.4	0.0	0.0	0.0	0.0

Works Cited

Addicott, John F. 1978. Competition for mutualists: aphids and ants. *Canadian Journal of Zool.* 56:2093–2096.

______. 1979. A multispecies aphid-ant association: density dependence and species-specific effects. *Canadian Journal of Zoology* 57(3):558–569.

______. 1981. Stability properties of 2-species models of mutualism: simulation studies. *Oecologica* 49(1):42–49.

______. 1984. Mutualistic interactions in population and community processes. In *A New Ecology: novel approaches to interactive systems*, edited by Peter W. Price, C. N. Slobodchikoff, and W. S. Gaud. John Wiley, New York. Pp. 437–456.

______. 1986. On the population consequences of mutualism. In *Community Ecology*, edited by Jared Diamond and Ted J. Case. Harper and Row, New York. Pp. 425–436.

Addicott, J. F., and H. I. Freedman. 1984. On the structure and stability of mutualistic systems: analysis of predator-prey and competition models as modified by the acton of a slow-growing mutualist. *Theoretical Population Biology* 26:320–339.

Ammerman, A., and L. L. Cavalli-Sforza. 1973. A population model for the diffusion of early farming in Europe. In *The Explanation of Culture Change: Models in Explanation*, edited by Colin Renfrew. Duckworth Press, London. Pp. 343–359.

Asch, David, and Nancy Asch. 1977. Chenopod as cultigen: a re-evaluation of some prehistoric collections from eastern North America. *Mid-Continental Journal of Archaeology* 2:3–45.

Aufdermauer, J., B. Dieckmann, and B. Fritsch. 1986. Die Untersuchung in einer bandkeramischen Siedlung bei Singen am Hohentwiel, Kreis Konstanz. In *Archäologische Ausgrabungen in Baden-Württemberg 1985*, edited by Dieter Plank. Konrad Theiss Verlag, Stuttgart. Pp. 51–53.

Aykroyd, W. R., and Joyce Doughty. 1970. *Wheat in Human Nutrition.* Food and Agriculture Organization United Nations, Rome.

Axelrod, Robert, and W. D. Hamilton. 1981. The evolution of cooperation. *Science* 211(27):1390–1396.

Works Cited

Bagniewski, Zbignew. 1981. Das Problem der Koexistenz mesolithischer und neolithischer Gesellschaften im Südteil des mitteleuropäischen Flachlandes. In *Mesolithikum in Europa*, edited by B. Gramsch. Veröffentlichungen des Museums für Ur- und Frühgeschichte Potsdam 14/15:113–119.

Bahuchet Serge, and Henri Guillaume. 1982. Aka-farmer relations in the northwest Congo Basin. In *Politics and History in Band Societies*, edited by E. Leacock and R. Lee. Cambridge University Press, Cambridge. Pp. 189–213.

Bakels, Corrie C. 1978. Four Linearbandkeramik settlements and their environment: a paleoecological study of Sittard, Stein, Elsloo, and Hienheim. *Analecta Praehistorica Leidensia* 11.

______. 1982. Zum wirtschaftlichen Nutzungsraum einer bandkeramischen Siedlung. In *Siedlungen der Kultur mit Linearkeramik in Europa*, edited by Juraj Pavúk. Archäologisches Institut der Slowakischen Akademie der Wissenschaften, Nitra. Pp. 9–16.

______. 1984. Carbonized seeds from Northern France. *Analecta Praehistorica Leidensia* 17:1–27.

Bakels, C. C., and R. Rousselle. 1985. Restes botaniques et agriculture du neolithique ancien en Belgique et aux Pays-Bas. *Helinium* 25:37–57.

Bandi, H.-G. 1963. *Birsmatten-Basisgrotte: eine mittelsteinzeitliche Fundstelle im unteren Birstal.* Acta Bernensia I. Stämpfli & Cie, Bern.

Barth, Frederik. 1956. Ecologic Relationships of Ethnic Groups in Swat, North Palistan. *American Anthropologist* 58:1079–1089.

______ (ed.). 1969. *Ethnic Groups and Boundaries.* Little Brown and Co., Boston.

Berlekamp, Hansdieter. 1977. Spätmesolithikum oder Altneolithikum? In *Archäologie als Geschichtswissenschaft.* Schriften zur Ur- und Frühgeschichte 30. Akademie Verlag, Berlin. Pp. 87–101.

Bertsch, Karl. 1931. Palaeobotanische Monographie des Federseeriedes. *Bibliotheca Botanica* 103:1–27.

______. 1954. Vom neolithischen Feldbau auf der Schwäbischen Alb. *Berichte der Deutschen Botanischen Gesellschaft* 67:8–10.

Bicker, F.-K. 1933. Mesolithisch-neolithische Kulturverbindungen in Mitteldeutschland? *Mannus* 25:249–252.

Binford, Lewis R. 1980. Willow smoke and dog's tails: hunter-gatherer settlement systems and archaeological site formation. *American Antiquity* 45(1):4–21.

Biswell, H. H., and M. D. Hoover. 1945. Appalachian hardwood trees browsed by cattle. *Journal of Forestry* 43:675–676.

Blackburn, Roderic. 1982. In the land of milk and honey: Okiek adaptations to their forests and neighbours. In *Politics and History in Band Societies*, edited by E. Leacock and R. Lee. Cambridge University Press, Cambridge. Pp. 283–306.

Works Cited

Boessneck, J., J.-P. Jéquier, and H. R. Stampfli. 1963. *Die Tierreste.* Acta Bernensia II Band 3.

Bogucki, Peter I. 1979. Tactical and strategic settlements in the Early Neolithic of Lowland Poland. *Journal of Anthropological Research* 35(2):236–242.

______. 1982. *Early Neolithic Subsistence and Settlement in the Polish Lowlands.* British Archaeological Reports International Series 150. British Archaeological Reports, Oxford.

______. 1984. Ceramic sieves of the Linear Pottery Culture and their economic importance. *Oxford Journal of Archaeology* 3(1):15–30.

Boucher, Douglas H. 1979. Seed predation and dispersal by animals in a tropical dry forest. PhD dissertation, University of Michigan, Ann Arbor, Michigan.

______. 1981. Seed predation by mammals and forest dominance by *Quercus oleoides*, a tropical lowland oak. *Oecologica* 49:409–414.

Boucher, D., S. Jones, and K. Keeler. 1982. The ecology of mutualism. *Annual Review of Ecology and Systematics* 13:315–347.

Bourke, A. 1984. Impact of climatic fluctuations on European agriculture. In *The Climate of Europe: past, present, and future*, edited by Hermann Flohn and Roberto Fantechi. Reidel, Dordrecht. Pp. 269–315.

Braun, David. 1977. Middle Woodland-Early Woodland Social Change in the Prehistoric Central Midwestern U.S. University Microfilms, Ann Arbor, MI.

Brenchley, W. E., and K. Warington. 1917. The effect of weeds upon cereal crops. *New Phytologist* 16:53–76.

______. 1930. The weed seed population of arable soil. I. Numerical estimation of viable seeds and observations on their natural dormancy. *Journal of Ecology* 18: 234–271.

Briggs, D. E. 1978. *Barley.* Chapman and Hall, London.

Brookfield, H. C., and P. Brown. 1963. *Struggle for Land.* Oxford University Press, Melbourne.

Brunnacker, K., W. v. Königswald, W. Rähle, F. Schweingruber, W. Taute, and W. Wile. 1979. Der Übergang vom Pleistozän zum Holozän in der Burghöhle von Deitfurt bei Sigmaringen. *Kölner Jahrbuch* 15:86–160.

Bryant, Larry D., and Chris Maser. 1982. Classification and distribution of elk in North America: ecology and management. *Elk of North America: ecology and management*, edited by J. W. Thomas and D. E. Toweill. United States Department of Agriculture, Forest Service. Stackpole Books, Boston. Pp. 1–60.

Buttler W., and W. Haberey. 1936. *Die bandkeramische Ansiedlung bei Köln-Lindental.* De Gruyter and Company, Berlin.

Works Cited

Campbell, John K. 1964. *Honor, Family, and Patronage.* Clarendon Press, Oxford.

Carleton, M. A. 1901. *Emmer, a grain for the semi-arid regions.* United States Department of Agriculture Farmers' Bulletin No. 139.

Carneiro, Robert L. 1961. Slash and burn cultivation among the Kikuru and its implications for cultural developments in the Amazon Basin. In *The Evolution of Horticultural Systems in Native South America: causes and consequences*, edited by Johannes Wilbert. Sociedad de Ciencias Naturales La Salle, Caracas. Pp. 67–80.

Case, T. J., and M. E. Gilpin. 1974. Interference competition and niche theory. *Proceedings of the National Academy of Sciences* 71:3073–3077.

Cavalli-Sforza, L. L. 1983. The transition to agriculture and some of its consequenses. In *How Humans Adapt*, edited by Donald S. Ortner. Smithsonian Press, Washington D.C. Pp. 103–127.

Chang, Cynthia. 1982. Nomads without cattle: East African foragers in historical perspective. In *Politics and History in Band Societies*, edited by E. Leacock and R. Lee. Cambridge University Press, Cambridge. Pp. 269–282.

Childe, V. Gordon. 1929. *The Danube in Prehistory.* Oxford University Press, Oxford.

Clark, J. G. D. 1952. *Prehistoric Europe: the economic basis.* Stanford University Press, Stanford.

———. 1980. *Mesolithic Prelude.* Edinburgh University Press, Edinburgh.

Clason, A. 1972. Viehzucht, Jagt und Knochenindustrie der Pfynerkultur. Ms. on file Biologisches Archaeologisches Instituut. Rijksuniversiteit, Groningen, The Netherlands.

Clutton-Brock, T., F. Guinness, and S. Albon. 1982. *Red Deer: behavior and ecology of two sexes.* University of Chicago Press, Chicago.

Cody, M. L. 1971. Finch flocks in the Mojhave Desert. *Theoretical Population Biology* 2:142–58.

Conklin, Harold. 1954. An ethnoecological approach to shifting cultivation. *Transactions of the New York Academy of Sciences*, Series 2, 17:133–142.

———. 1957. *Hanunoo Agriculture: a report on an integral system of shifting cultivation in the Philippines.* Food and Agriculture Organization, Rome.

Cooke, G. W. 1976. Long-term fertilizer experiments in England: the significance of their results for agricultural science and for practical farming. *Annales Agronomiques* 27:503–536.

Cowan, C. Wesley. 1985. *From Foraging to Incipient Food Production: subsistence change and continuity on the Cumberland Plateau of eastern Kentucky.* University Microfilms, Ann Arbor, MI.

Cranstone, B. A. L. 1969. Animal husbandry: the evidence from ethnography. In *Domestication of Plants and Animals*, edited by Peter J. Ucko and G. W. Dimbleby. Duckworth Press, London. Pp. 231-247.

Dansgaard, W. 1984. Past climates and their relevance. In *The Climate of Europe: past, present, future*, edited by Hermann Flohn and Roberto Fantechi. Reidel, Dordrecht. Pp. 198–265.

Davidson, Stanley, R. Passmore, and J. Brock. 1979. *Human Nutrition and Dietetics*. Williams and Wilkins, Baltimore.

Dean, Antony. 1983. A simple model of mutualism. *American Naturalist* 121(3):409–416.

Dennell, Robin. 1983. *European Economic Prehistory: a new approach*. Academic Press, New York.

Dressler, Robert L. 1982. Biology of the orchid bees (Euglossini). *Annual Review of Ecology and Systematics* 13:373–394.

Dyson-Hudson, Rada, and N. Dyson-Hudson. 1970. The food production system of a semi-nomadic society: the Karimojong, Uganda. In *African Food Production Systems*, edited by Peter McLoughlin. Johns Hopkins Press, Baltimore. Pp. 91–123.

Eder, James E. 1984. The impact of subsistence change on mobility and settlement patterns in a tropical forest foraging economy: some implications for archaeology. *American Anthropologist* 86:837–853.

Ellenberg, Heinz. 1950. Unkraut-Gemeinschaften als Zeiger für Klima und Boden. Eugen Ulmer Verlag, Ludwigsburg.

_______. 1952. *Wiesen und Weiden und ihre standörtliche Bewertung*. Landwirtschaftliche Pflanzensoziologie 11.

_______. 1979 *Zeigerwerte der Gefässpflanzen Mitteleuropas*. Scripta Geobotanica 9. Second Edition.

_______. 1982. *Vegetation Mitteleuropas in ökologischer Sicht*. Third Edition. Eugen Ulmer Verlag, Stuttgart.

Ensminger, A. H., M. Ensminger, J. Konlande, and J. Robson. 1983. *Foods and Nutrition Encyclopedia*. Vols. 1 and 2. Pegasus Press, Clovis, CA.

Erviö, Leila-Riita. 1972. Growth of weeds in cereal populations. *Journal of the Scientific Agricultural Society of Finland* 44:19–28.

Evans-Pritchard, E. E. 1940. *The Nuer*. Oxford University Press, Oxford.

Food and Agriculture Organization. 1953. *Food Composition Tables*. Food and Agriculture Organization Nutritional Studies No. 3. United Nations, Rome.

Fabricus, L. J., and J. D. Nalewaja. 1968. Competition between wheat and wild buckwheat. *Weed Science* 16(2):204–208.

Fansa, Mamoun. 1985. Die Jungsteinzeit in Niedersachsen. In *Ausgrabungen in Niedersachsen: Archäologische Denkmalpflege 1979–1984*. Konrad Theiss Verlag, Stuttgart. Pp. 83–86.

Fansa, Mamoun, and Hartmut Thieme. 1985. Eine Siedlung und Befestigungsanlage der Bandkeramik auf dem "Nachtwiesen-Berg" bei Esbeck, Stadt Schöningen, Landkreis Helmstedt. In *Ausgrabungen in Niedersachsen: Archäologische Denkmalpflege 1979–1984*. Konrad Theiss Verlag, Stuttgart. Pp. 87–92.

Farrugia, J. P., R. Kuper, J. Lüing, and P. Stehli. 1973. Untersuchungen zur neolithischen Besiedlung der Aldenhovener Platte III. *Bonner Jahrbücher* 173:226–256.

Firbas, F. 1949. *Spät- und nacheiszeitliche Waldgeschichte Mitteleuropas nordlich der Alpen*. Fischer, Jena.

Flohn, Hermann, and Roberto Fantechi (eds.). 1984. *The Climate of Europe: past, present, future*. Reidel, Dordrecht.

Födisch, Hermann. 1961. Neolithische Silexgeräte aus Inzkofen, Ldkr. Freising. *Bayerische Vorgeschichtsblätter* 26:123–128.

Ford, Richard I. 1978. Gathering and gardening: trends and consequences of Hopewell subsistence strategies. In *Hopewell Archaeology*, edited by David S. Browse and N'omi Greber. Kent State University Press, Kent, Ohio. Pp. 234–238.

_______. 1979. Paleoethnobotany in American Archaeology. *Advances in Archaeological Method and Theory* 2:285–336.

Fox, Richard. 1969. Professional primitives. *Man in India* 49:139–60.

Freeman, Derek. 1970. *Report on the Iban*. Monographs on Social Anthropology No. 41. Althlone Press, London.

Frenzel, B. 1966. Climatic change in the Atlantic/Sub-Boreal Transition on the Northern Hemisphere: botanical evidence. In *Proceedings of the International Symposiuum of World Climate from 8000–0 B.C*, edited by J. S. Sawyer. Royal Meteorological Society, London. Pp. 99–123.

Fried, Morton. 1975. *The Notion of Tribe*. Cummings Press, Menlo Park.

Friesen, G., and L. Shebeski. 1960. Economic losses caused by weed competition in Manitoba grain fields. *Canadian Journal of Plant Science* 40:457–467.

Futuyma, D. J. 1979. *Evolutionary Biology*. Sinauer Associates, Sunderland, MA.

Georgia, A. 1938. *A Manual of Weeds*. Macmillan Co., New York.

Gersbach, E. 1956. Ein Harpunenbruchstück aus einer Grube der jüngeren Linearbandkeramik. *Germania* 34:266–70.

Geupel, Volkmar. 1981. Zum Verhältnis Spätmesolithikum-Frühneolithikum im mittleren Elbe-Saale-Gebiet. In *Mesolithikum in Europa*. Veröffentlichungen des Museums für Ur- und Frühgeschichte Potsdam 14/15:105–112.

Gill, N.T., and K.C. Vear. 1980. *Agricultural Botany* Vols. 1 and 2. Third Edition. Duckworth Press, London.

Gradmann, Robert. 1902. Beziehungen zwischen Pflanzengeographie und Siedlungsgeschichte. *Geographische Zeitschrift* 12:305–325.

Gramsch, Bernhard. 1971. Zum Problem des Übergangs vom Mesolithikum zum Neolithikum in Flachland zwischen Elbe und Oder. In *Evolution und Revolution im alten Orient und in Europa*, edited by F. Schlette. Akademie Verlag, Berlin. Pp. 127–144.

______ (ed.). 1981. *Mesolithikum in Europa.* Veröffentlichungen des Museums für Ur- und Frühgeschichte Potsdam 14/15:113–119.

Green, Stanton. 1977. *The Agricultural Colonization of Temperate Forest Habitats: an ecological model.* University Microfilms, Ann Arbor, MI.

Gregg, S. 1980. A material perspective of tropical rain forest hunter-gatherers: the Semang of Malaysia. In *The Archaeological Correlates of Hunter-Gatherer Societies.* Michigan Discussions in Anthropology 5:117–136. Department of Anthropology, University of Michigan, Ann Arbor, MI.

______. 1984. Die vorläufigen Ergebnisse der paläo-ethnobotanischen Untersuchungen der bandkeramischen Siedlungen bei Ulm-Eggingen. *Archäologica Venatoria Mitteilungsblatt* 7:25–33.

Griffin, P. Bion. 1984. Forager Resource and Land Use in the Humid Tropics: The Agta of Northeastern Luzon, the Philippines. In *Past and Present in Hunter Gatherer Studies*, edited by Carmel Schrire. Orlando: Academic Press. Pp. 95–121

Grigson, Caroline. 1981. Fauna. In *The Environment in British Prehistory*, edited by Ian Simmons and Michael Tooley. Pp. 191–199; 217–230.

______. 1982a. Porridge and pannage: pig husbandry in Neolithic Europe. In *Archaeological Aspects of Woodland Ecology*, edited by Martin Bell and Susan Limbrey. British Archaeological Reports International Series 146. British Archaeological Reports, Oxford. Pp. 297–312.

______.1982b. Sex and age determinations of some bones and teeth of domestic cattle: a review of the literature. In *Aging and Sexing Animal Bones from Archaeological Sites*, edited by Bob Wilson, Caroline Grigson, and Sebastian Payne. British Archaeological Reports International Series 109. British Archaeological Reports, Oxford.

Groenman-van Wateringe, W. 1971. Hecken im Westeuropäischen Frühneolithikum. *Berichten van de Rijksdienst voor Het Oudheidkundig Bodemonderzoek* 150(20–21):295–299.

Groenman-van Wateringe, W., and W. van Wateringe. 1979. The origin of crop weed communities composed of summer annuals. *Vegetatio* 41:57–59.

Gross, Daniel K. 1983. Village movement in relation to resources in Amazonia. In *Adaptive Responses to Native Amazonians*, edited by Raymond B.ames and William T. Vickers. Academic Press, New York.

Gross, W. L. 1924. The vitality of buried seeds. *Journal of Agricultural Research* 29(7):349–362.

Gruenhagen, R., and J. Nalewaja. 1969. Competition between flax and wild buckwheat. *Weed Science* 17(3):380–384.

Hahn, Joachim. 1983. Eiszeitliche Jäger zwischen 35,000 und 15,000 vor heute. In *Urgeschichte in Baden-Württemberg*, edited by H. Müller-Beck. Konrad Theiss Verlag, Stuttgart. Pp. 273–330.

Hahn, J., H. Müller-Beck, and W. Taute. 1973. *Eiszeithöhlen im Lonetal.* Verlag Müller & Gräff, Stuttgart.

Hall, A. D. 1917. *The Book of Rothamstead Experiments.* Second Edition. (revised by E. J. Russell). J. Murray, London.

Hammond, Frederik W. 1981. The colonization of Europe: the analysis of settlement process. In *Pattern of the Past: studies in honour of David Clarke*, edited by I. Hodder, G. Isaac, and N. Hammond. Cambridge University Press, Cambridge. Pp. 211–248.

Harako, Reizo. 1976. The Mbuti as hunters—a study of ecological anthropology of the Mbuti Pygmies (I). *Koyoto University African Studies* 10:37–99.

Harm, G. W., and F. D. Keim. 1934. The percentage and viability of weed seeds recovered in the feces of farm animals and their longevity when buried in manure. *American Society of Agronomy Journal* 26:762–767.

Harris, Michael. 1980. *Heating with Wood.* Citadel Press, Secaucus, NJ.

Hart, John A. 1978. From subsistence to market: a case study of the Mbuti net hunters. *Human Ecology* 6(3):325–353.

———. 1979. *Nomadic Hunters and Village Cultivators: a study of subsistence interdependence in the Ituri Forest of Zaire.* M.A. Thesis, Department of Geography, Michigan State University, East Lansing, MI.

Hart, T. B., and J. A. Hart. 1986. The ecological basis of hunter-gatherer subsistence in African rainforests: the Mbuti of Eastern Zaire. *Human Ecology* 14:29–56.

Hawtin, G. C., K. B. Signh, and M. C. Saxena. 1980. Some recent developments in the understanding and improvement of Cicer and Lens. In *Advances in Legume Science*, edited by R. J. Summerfield and A. H. Bunting. Royal Botanic Gardens, Kew, England. Pp. 50–76.

Headland, Thomas. 1978. Cultural ecology, ethnicity, and the Negritos of Northeastern Luzon. *Asian Perspectives* 21:127–139.

Heer, Oswald. 1865. Die Pflanzen der Pfahlbauten. *Neujahrsblatt der Naturforschenden Gesellschaft Zürich für das Jahr 1866* 68:1–54.

Hegmon, Michelle. 1985. Exchange in social integration and subsistence risk: a computer simulation. Manuscript on file. Museum of Anthropology, Ann Arbor, MI.

Works Cited

Heidenreich, Conrad. 1971. *Huronia: a history and geography of the Huron Indians, 1600–1650.* McClelland and Stewart, Toronto.

_______. 1978. Huron. In *Northeast*, edited by Bruce G. Trigger. Handbook of North American Indians, vol. 15, William G. Sturdevant, general editor. Smithsonian Institution Press, Washington DC. Pp. 368–389.

Heiligmann, Jörg. 1983. Archäologische Untersuchungen in einer bandkeramischen Siedlung bei Ulm-Eggingen, Stadtkreis Ulm. In *Archäologische Ausgrabungen in Baden-Württemberg 1982*, edited by Dieter Plank. Konrad Theiss Verlag, Stuttgart. Pp. 27–29.

Helbaek, Hans. 1960. Comment on *Chenopodium album* as a food plant in prehistory. *Bericht des Geobotanischen Instituts der Eidgenössischen Technischen Hochschule* 31.

Herweijer, C. H., and L. F. den Houter. 1970. Poisoning in sheep caused by *Chenopodium album. Tijdschrift voor Diergeneeskunde* 95:1134–1136.

Higham, Charles. 1966. *Stock-rearing in Prehistoric Europe.* PhD thesis, Cambridge University.

Hillman, Gordon C. 1984. Interpretation of archaeological plant remains: the application of ethnographic models from Turkey. In *Plants and Man*, edited by W. van Zeist and W. A. Casparie. Balkema: Rotterdam. Pp. 1–43.

Hitchcock, Robert K. 1982. *The Ethnoarchaeology of Sedentism: mobility strategies and site structure among foraging and food producing populations in the Eastern Kahalari, Botswana.* University Microfilms: Ann Arbor, MI.

Hoffman, Walter J. 1896. The Menomini Indians. *14th Annual Report of the Bureau of Ethnology. Part I.* Washington DC.

Hoffmann, Walther. 1961. Lein, *Leinum usitatissimum* L. In *Handbuch der Pflanzenzüchtung.* Parey Verlag, Berlin. Pp. 264–366.

Holm, LeRoy, D. Plucknen, and J. Pancho. 1977. *The World's Worst Weeds.* University Press of Hawaii, Honolulu.

Howe, Henry F. 1984. Constraints on the evolution of mutualisms. *American Naturalist* 123(6):764–777.

Howe, Henry F., and Judith Smallwood. 1982. Ecology of seed dispersal. *Annual Review of Ecology and Systematics* 13:201–228.

Howell, John M. 1983. *Settlement and Economy in Neolithic Northern France.* British Archaeological Reports International Series 157. British Archaeological Reports, Oxford.

Hume, L. 1982. The long-term effects of fertilizer application and three-year rotation on weed communities after 21–22 years at Indian Head, Saskatchewan. *Canadian Journal of Plant Science* 62:741–750.

Works Cited

Hume, L., J. Martinez, and K. Best. 1983. The biology of Canadian weeds. 60. *Polygonum Convolvulus* L. *Canadian Journal of Plant Science* 63:57–65.

Huston, J. E. 1979. Forage utilization and nutrient requirements of the Goat. *Journal of Dairy Science* 61: 988–993.

Hutterer, Karl L. 1976. An evolutionary approach to the Southeast Asian cultural sequence. *Current Anthropology* 17(2):221–241.

Iverson, Johs. 1973. The development of Denmark's nature since the last Glacial. *Danmarks Geologiske Undersøgelse* 5 (7–8):7–126.

Izikowitz, Karl G. 1979. *Lamet: hill peasants in French Indochina.* AMS Press, New York.

Jacomet, Stefanie, and Jörg Schibler. 1985. Die Nahrungsversorgungen eines jungsteinzeitlichen Pfynerdorfes am unteren Zürichsee. *Archäologie der Schweiz* 8:125–140.

Jacomet, Stefanie, and Helmut Schlichtherle. 1984. Der kleine Pfahlbauweizen Oswald Heer's – Neue Untersuchungen zur Morphologie neolithischer Nacktweizen-Ähren. In *Plants and Ancient Man*, edited by W. van Zeist and W. Casparie. Balkema, Rotterdam. Pp. 153–177.

Janson, C. H., J. Terborgh, and L. H. Emmons. 1981. Non-flying mammals are pollinating agents in the Amazonian forest. *Biotropica* 13(2): Supp. 1–6.

Jochim, Michael 1976. *Hunter-Gatherer Subsistence and Settlement: a predictive model.* Academic Press, New York.

——— (ed.). In press. *Der Mittelsteinzeitliche Lagerplatz Henauhof N.W. am Federsee.* Konrad Theiss Verlag, Stuttgart.

Jochim Michael, and Susan A. Gregg. 1984. Mittelsteinzeitliche Forschungen im Federseegebiet. In *Archäologische Ausgrabungen in Baden-Württemberg 1983*, edited by D. Plank. Konrad Theiss Verlag, Stuttgart. Pp. 84–89.

Johnson, Gregory A. 1982. Organizational structure and scalar stress. In *Theory and Explanation in Archaeology*, edited by C. Renfrew, M. Rowlands, and B. Seagraves. Academic Press, New York. Pp. 389–421.

Johnston, A. E., and G. E. Mattingly. 1976. Experiments in the continuous growth of arable crops at Rothamstead and Woburn experimental stations: effects of treatments on crop yields and soil analyses and recent modifications in purpose and design. *Annales Agronomiques* 27:927–956.

Jordano, Pedro. 1987. Patterns of mutualistic interactions in pollen and seed dispersal: connectance, dependence, asymmetries, and coevolution. *American Naturalist* 129(5):657–677.

Keegan, William F. 1986. The optimal foraging analysis of horticultural production. *American Anthropologist* 88(1):92–108.

Keeler, Kathleen H. 1981. A model of selection for facultative, nonsymbiotic mutualism. *American Naturalist* 118:488–98.

Keene, Arthur. 1981. *Prehistoric Foraging in a Temperate Forest: a linear program model.* Academic Press, New York.

Kelley, J. Charles. 1955. Juan Sabeata and diffusion in aboriginal Texas. *American Anthropologist* 57:981-995.

______. 1986 *Jumano and Patarabueye: relations at La Junta de los Rios.* Anthropological Papers No. 77. Museum of Anthropology, Ann Arbor, MI.

Kelly, Raymond. 1977. *Etoro Social Structure: a study in structural contradiction.* University of Michigan Press, Ann Arbor, MI.

Kelly, Robert L. 1985. *Hunter-Gatherer Mobility and Sedentism: a Great Basin study.* University Microfilms, Ann Arbor, MI.

Kind, Claus-Joachim. 1984. Die Ausgrabungen 1983 in der bandkeramischen Sieldung von Ulm-Eggingen, Alb-Donau-Kreis. In *Archäologische Ausgrabungen in Baden-Württemberg 1982*, edited by Dieter Plank. Konrad Theiss Verlag, Stuttgart. Pp. 43–47.

______. 1986. Die abschliessende Grabungskampagne 1985 in Ulm-Eggingen, Stadtkreis Ulm. In *Archäologische Ausgrabungen in Baden-Württemberg 1986*, edited by Dieter Plank. Konrad Theiss Verlag, Stuttgart. Pp. 45–50.

Kislev, M. E. 1984. Botanical evidence for ancient naked wheats in the Near East. In *Plants and Ancient Man*, edited by W. van Zeist and W. A. Casparie. Balkema, Rotterdam. Pp. 141–153.

Klingmann, Glenn C. 1961. *Weed Control as a Science.* John Wiley, New York.

Knörzer, K.-H. 1967. Die Roggentrespe (*Bromus secalinus* L.) als prähistorische Nutzpflanzen. *Archaeo-Physika* 2:30–37.

______. 1971a. Urgeschichtliche Unkräuter im Rheinland: ein Beitrag zur Entstehungsgeschichte der Segetalgesellschaften. *Vegetatio* 23(1–2):89–111.

______. 1971b. Pflanzliche Grossreste aus der rössenerzeitlichen Siedlung bei Langweiler, Kreis Jülich. *Bonner Jahrbücher* 171:9–33.

______. 1971c. Genutzte Wildpflanzen in vorgeschichtlicher Zeit. *Bonner Jahrbücher* 171:1–8.

______. 1973. Der bandkeramische Siedlungsplatz Langweiler 2: pflanzliche Grossreste. *Rheinische Ausgrabungen* 13: 139–152.

______. 1979. Über den Wandel der angebauten Körnerfrüchte und ihrer Unkrautvegetation auf einer niederrheinischen Lössfläche seit dem Frühneolithikum. *Archaeo-Physika* 8:147–163.

______. 1980. Pflanzliche Grossreste des bandkeramischen Siedlungsplatzes Wanlo (Stadt Mönchen-Gladbach). *Archeo-Physika* 7:7–20.

Kormso, Emil. 1930. *Unkräuter im Ackerbau der Neuzeit.* Springer Verlag, Berlin.

Kossack, Georg and Hans Schmeidl. 1974. Vorneolithischer getreidebau im bayerischen Alpenvorland. *Jahresbericht der Bayerischen Bodendenkmalpflege* 15/16:7–23.

Koyama, Shuzo, and David H. Thomas (eds.). 1979. *Affluent Foragers*. Senri Ethnological Series 11.

Kruk, Janusz. 1980. *The Neolithic Settlement of Southern Poland*. British Archaeological Reports International Series 93. British Archaeological Reports, Oxford.

Kuper, R., H. Löhr, J. Lüning, P. Stehli, and A. Zimmerman. 1977. Der bandkeramische Siedlungsplatz Langweiler 9. *Rheinische Ausgrabungen* 18.

Lamb, H. H. 1977. *Climate Present, Past and Future*. Methuen, New York.

Laursen, Franz. 1971. Studies of weed competition in barley. *Yearbook of Royal Veterinary and Agriculture College* 1971:201–222.

Lawton, Harry W., and Lowell J. Bean. 1968. A preliminary reconstruction of aboriginal agricultural techniques among the Cahuilla. *Indian Historian* 1:18–24.

Leacock, Eleanor, and Richard B. Lee (eds.) 1982. *Politics and Hierarchy in Band Societies*. Cambridge University Press, Cambridge.

Lee, Richard, and Irven DeVore. 1968. *Man the Hunter*. Aldine, Chicago.

Lewis, D. H. 1973. The relevance of symbiosis to taxonomy and ecology, with particular reference to mutualistic symbioses and exploitation of marginal habitats. In *Taxonomy and Ecology*, edited by V. H. Heywood. Academic Press, New York. Pp. 151–172.

Lockwood, William G. 1975. *European Moslems*. Academic Press, New York.

Loomis, R. S. 1978. Ecological dimensions of medieval agrarian systems: an ecologist responds. *Agricultural History* 52:478–483.

Lüning, Jens. 1976. Un nouveau modèle de l'habitat du Néolithique ancien. Ninth Congress UISPP Nice. CNRS, Paris

_______. 1982. Siedlung und Siedlungslandschaft in bandkeramischer und Rössner Zeit. *Offa* 39:9.

Lüning, Jens, and Jutta Meurers-Balke. 1980. Experimenteller Getreideanbau im Hambacher Forst, Gemeinde Elsdorf, Kr. Bergheim/Rheinland. *Bonner Jahrbücher* 180:305–344.

Mackenzie, David. 1980. *Goat Husbandry*. Faber and Faber, London.

Martin, John H., and S. C. Salmon. 1953. The rusts of wheat, oats, barley, and rye. *Yearbook of Agriculture* 1953:329–343.

Maurizio, A. 1927. *Die Geschichte unserer Pflanzennahrung*. Parey Verlag, Berlin.

Mausch, Anna, and Karl-Heinz Ziessow. 1985. Reconstructing linear culture houses: theoretical and practical contributions. *Helinium* 25:58–93.

May, Robert. 1981. Models for two interacting populations. In *Theoretical Ecology*. Sinauer Associates, Sunderland, MA. Pp. 78–105.

Mazálek, M. 1953. Zur Frage der Beziehungen zwischen Mesolithikum und Neolithikum. *Anthropozoikum* 3:224–234.

McMahan, Craig. 1968. Comparative food habits of deer and three classes of livestock. *Journal of Wildlife Ecology* 28:798–808.

Menke, Manfred. 1978. Zum Frühneolithikum zwischen Jura und Alpenrand. *Germania* 56:24–52.

Meyer-Christian, W. 1976. Die Y-pfostenstellung in Häusern der Älteren Linearbandkeramik. *Bonner Jahrbücher* 176:1–25.

Milisauskas, Sarunas. 1976. *Archaeological Investigations on the Linear Culture Village of Olszanica*. Ossolineum, Wroclaw

______. 1977. Adaptations of the early neolithic farmers in Central Europe. In *For the Director: research essays in Honor of James B. Griffin*. Anthropological Papers 61. Museum of Anthropology, Ann Arbor, MI. Pp. 295–316.

______. 1978. *European Prehistory*. Academic Press, New York.

Milojčić, V. 1952. Die frühsten Ackerbauer Mitteleuropas. *Germania* 30:313–318.

Milton, Katherine. 1984. Protein and carbohydrate resources of the Maku Indians of Northwestern Amazonia. *American Anthropologist* 86(1):7–27.

Modderman, Pieter J. R. 1970. Linearbandkeramik aus Elsloo und Stein. *Analecta Praehistorica Leidensia* 3.

______. 1971. Bandkeramiker und Wanderbauerntum. *Archäologisches Korrespondenzblatt* 1:7–9.

Morris, Brian. 1982. The family, group structuring and trade among South Indian hunter-gatherers. In *Politics and History in Band Societies*, edited by E. Leacock and R. Lee. Cambridge University Press, Cambridge. Pp. 171–188.

Muenscher, Walter C. 1980. *Weeds*. Second Edition. Cornell University Press, Ithaca.

Müller, Hans-Hermann. 1964. *Die Haustiere der mitteldeutschen Bandkeramiker*. Deutsche Akademie der Wissenschaften zu Berlin, Schriften der Sektion für Vor- und Frühgeschichte, Band 17. Akademie Verlag, Berlin.

Müller, Inge. 1947. Der pollenanalytische Nachweis der menschlichen Besiedlung im Federsee- und Bodesnseegebiet. *Planta* 35:70–87.

Müller-Beck, H. 1983. Die Späte Mittelsteinzeit. In *Urgeschichte in Baden-Württemberg*, edited by H. Müller-Beck. Konrad Theiss Verlag, Stuttgart. Pp. 393–404.

National Academy of Sciences. 1979. *Recommended Dietary Allowances*, Ninth Edition. National Academy of Sciences, Washington DC.

Needham, R. 1962. Genealogy and category in Wikmunkan Society. *Ethnology* 1:223–64.

Nelson, J. R. 1982. Relationships of elk and other large herbivores. In *Elk of North America: ecology and management*, edited by J. Thomas and D. E. Toweill. United States Department of Agriculture, Forest Service. Stackpole Books, Boston. Pp. 415–441.

Netting, Robert. 1968. *Hill Farmers of Nigeria: cultural ecology of the Kofayr of the Jos Plateau*. American Ethnological Society Monograph 46. University of Washington Press, Seattle.

______. 1981. *Balancing on an Alp*. Cambridge University Press, Cambridge.

Neumann, A. L., and R. R. Snapp. 1969. *Beef Cattle*. John Wiley, New York.

Neuweiler, E. 1905. Die prähistorischen Pflanzenreste Mitteleuropas mit besonderer Berücksichtigung der schweizerischen Funde. *Vierteljahresschrift der Naturforschenden Gesellschaft Zürich* 50:23–134.

NRC. 1975. *Nutrient Requirements of Sheep*. Nutrient Requirements of Domestic Animals No. 5. National Academy of Sciences: Washington DC.

______. 1981. *Nutrient Requirements of Goats*. Nutrient Requirements of Domestic Animals No. 15. National Academy of Sciences, Washington DC.

______. 1984. *Nutrient Requirements of Beef Cattle*. Sixth Edition. National Research Council, National Academy Press, Washington DC.

Oberdorfer, Erich. 1979. *Pflanzensoziologische Exkursions Flora*. Eugen Ulmer Verlag, Stuttgart.

Odum, Eugene. 1979. *Fundamentals of ecology*. Second Edition. Saunders, Philadelphia.

OECD. 1976. *Milk and Milk Product Balances in OECD Member Countries 1961–1974*. Organization for Economic Co-Operation and Development, Paris.

______. 1984. *Milk and Milk Product Balances in OECD Countries 1974–1982*. Organization for Economic Co-operation and Development, Paris.

Ogg, Alex G., and Jean H. Dawson. 1984. Time of emergence of eight weed species. *Weed Science* 32:327–335.

Owen, Robert C. 1965. The patrilocal band: a linguistically and culturally social unit. *American Anthropologist* 67(3):675–690.

Palti, J. and C. Paludan-Müller.1981. *Cultural Practices and Infectious Crop Diseases*. Springer Verlag, Berlin. Pp. 120–157.

Parsons, James J. 1962. The acorn-hog economy of the oakwoodlands of southwestern Spain. *Geographical Review* 52(2):211–235.

Patrick, E. F., and W. L. Webb. 1960. An evaluation of three age determination criteria in live beavers. *Journal of Wildlife Management* 24:37–44.

Pellett, P. L., and W. Shadarevian. 1970. *Food Composition Tables for Use in the Middle East.* American University, Beruit.

Penistan, M. J. 1974. Growing Oak. In *The British Oak*, edited by M. G. Morris and F. H. Perring. The Botanical Society of the British Isles: Berkshire, England. Pp. 98–113.

Percival, John. 1921. *The Wheat Plant.* Duckworth: London.

Perry, Tilden Wayne. 1984. *Animal Life-Cycle Feeding and Nutrition.* Academic Press, New York.

Peters, E. 1934. Das Mesolithikum der oberen Donau. *Germania* 18:81–89.

Peterson, J. T. 1978a. *The Ecology of Social Boundaries: Agta foragers of the Philippines.* Illinois Studies in Anthropology 11. University of Illinois Press, Urbana.

_______. 1978b. Hunter-gatherer/farmer exchange. *American Anthropologist* 80:335–351.

Peterson, R. F. 1965. *Wheat: botany, cultivation and utilization.* Leonard Hill, London.

Pianka, Eric R. 1978. *Evolutionary Ecology.* Second Edition. Harper and Row, New York.

Pierce, N. E., and P. S. Mead. 1981. Parasitoids as selective agents in symbiosis between lycaenid butterfly larvae and ants. *Science* 211:1185–87.

Pookajorn, Surin. 1982. *The Phi Tong Luang: a hunter-gatherer group in Thailand.* Faculty of Archaeology, Silpakorn Univerity, Bankok, Thailand. Vol. 3 No. 1.

Quitta, H. 1960. Zur Frage der ältesten Bandkeramik in Mitteleuropa. *Prähistorische Zeitschrift* 38 1–38, 153–188.

Redding, Richard W. 1981. *Decision Making in Subsistence Herding of Sheep and Goats in the Middle East.* University Microfilms, Ann Arbor, MI.

Regal, Philip. 1982. Pollination by wind and animals: ecology of geographic patterns. *Annual Review of Ecology and Systematics* 13:497-527

Reinerth, H. 1928. Oberschwäbisches Mesolithikum. *Nachrichten der Deutschen Anthropologischen Gesellschaft* 3:77–82.

_______. 1936. *Der Federsee als Siedlungsland des Vorzeitmenschen.*

Reynolds, Peter J. 1984. Deadstock and livestock. In *Farming Practices in British Prehistory*, edited by Roger Mercer. Edinburgh University Press, Edinburgh. Pp. 97–122.

Ricklefs, Robert E. 1979. *Ecology.* Second Edition. Chiron Press, New York.

Ridley, H. 1930. *The Dispersal of Plants Throughout the World.* L. Reve & Co., Ashford, England.

Riek, G. 1934. *Die Eiszeitjägerstation am Vogelherd. Bd. I: Die Kulturen.* Heine, Tübingen.

Risch. S., and D. H. Boucher. 1977. What ecologists look for. *Ecological Society of America Bulletin* 57:8.

Robbins, Wilfred. 1952. *Weed Control: a textbook and manual.* McGraw-Hill, New York.

Roberts, H. A. 1982. *Weed Control Handbook: principles.* Blackwell, London.

Roughgarden, Jonathan. 1975. Evolution of marine symbiosis: a simple cost-benefit model. *Ecology* 56:1201–1208.

Rowley-Conway, Peter. 1984. The laziness of the short-distance hunter: the origins of agriculture in western Denmark. *Journal of Anthropological Archaeology* 3:1–25.

Ryder, Michael L. 1983. *Sheep and Man.* Duckworth, London.

Ruthenberg, Hans. 1980. *Farming Systems in the Tropics.* Clarendon Press, Oxford.

Sahlins, Marshall. 1968. *Tribesmen.* Prentice-Hall, Englewood Cliffs, NJ.

_______. 1972. *Stone Age Economics.* Aldine, Chicago.

Sakamoto, Clarence M. 1981. The technology of crop/weather modeling. In *Food-Climate Interactions*, edited by Wilfred Bach, Jürgen Pankrath, and Stephen Schneider. Reidel, Dordrecht. Pp. 383–398.

Sangmeister, Edward. 1951. Zum Charakter der bandkeramischen Siedlung. *Bericht der Römish-Germanischen Kommission* 33:89–95

_______. 1983. Die ersten Bauern. In *Urgeschichte in Baden-Württemberg*, edited by H. Müller-Beck. Thiess Verlag, Stuttgart. Pp. 429–473.

Scheffer, Charles. 1972. Notes on agronomy with special reference to wheat, barley, sorghum, and the millets. Unpublished manuscript on file at the Museum of Anthropology, Ann Arbor, MI.

Scheffrahn, Wolfgang. 1969. Die menschlichen Populationen. *Ur- und Frühgeschichtliche Archäologie der Schweiz* 2:33–41.

Schlichtherle, Helmut. 1984. Die Sondagen des "Projekts Bodensee-Oberschwaben" als Vorbereitung neuer siedlungsarchäologischer Forschungen in den Seen und Mooren Südwestdeutschlands. *Berichte zu Ufer- und Moorsiedlungen Südwestdeutschlands* 1:9–37.

Schlippe, Pierre de. 1956. *Shifting Cultivation in Africa.* Routledge and Kegan Paul, London.

Schmidt, Klaus. 1984. Zwei neue Kareten zur nacheiszeitlichen Besiedlungsgeschichte des Federseebeckens. *Berichte zu Ufer- und Moorsiedlungen Südwestdeutschlands* 1:101–115.

Schmidt, R. R. 1936. *Jungsteinzeit-Siedlungen im Federseemoor.* Ferdinand Enke Verlag, Stuttgart.

Schoener, Thomas. 1977. Competition and the niche. In *BIology of the Reptiles*, edited by D. W. Tinkle and H. Gans. Academic Press, New York.

Schönweiss, W., and H. Werner. 1974. Mesolithische Wohnlagen von Sarching, Ldkr. Regensburg. *Bayerische Vorgeschbl.* 39:1–29.

Schütrumpf, R. 1968. Die neolithischen Siedlungen von Ehrenstein bei Ulm, Aichbühl und Riedschachen im Federseemoor im Lichte moderner Pollenanalyse. In *Das jungsteinzeitliche Dorf Ehrenstein* (Kreis Ulm). Veröffentlichungen des Staatlichen Amtes für Denkmalpflege Stuttgart, Reihe A, 10/11:79–104.

Scudder, Thayer. 1962. *The Ecology of the Gwembe Tonga.* Manchester University Press, Manchester.

Sen Gupta, P. N. 1980. Food consumption and nutrition of regional tribes of India. *Ecology of Food and Nutrition* 9:39–108.

Service, Elman R. 1962. *Primitive Social Organization.* Random House, New York.

Shaw, M. W. 1974. The reproductive characteristics of oak. In *The British Oak*, edited by M. G. Morris and F. H. Perring. The Botanical Society of the British Isles, Berkshire, England. Pp. 162–181.

Sherratt, A. 1981. Plough and pastoralism: aspects of the secondary products revolution. In *Pattern of the Past: studies in honour of David Clarke*, edited by I. Hodder, G. Isaac, and N. Hammond. Cambridge University Press, Cambridge. Pp. 261–305.

Sielmann, B. 1971. Der Einfluss der Umwelt auf die neolithische Besiedlung Südwestdeutschlands unter besonderer Berücksichtigung der Verhältnisse am nördlichen Oberrhein. *Acta Praehistorica et Archaeologica* 2:65–197.

Sinha, D. P. 1972. The Birhors. In *Hunters and Gatherers Today*, edited by M. G. Bicchieri. Holt, Rinehart and Winston, New York. Pp. 67–81.

Simms, Steven R. 1984. *Aboriginal Great Basin Foraging Strategies: an evolutionary perspective.* University Microfilms, Ann Arbor, MI.

Smiley, F. E. 1981. The Birhor: material correlates of hunter-gatherer / farmer exchange. In *The Archaeological Correlates of Hunter-Gatherer Societies: studies from the ethnographic record*, Michigan Discussions in Anthropology 5:117–136. Department of Anthropology, University of Michigan, Ann Arbor, MI.

Smith, N. G. 1968. The advantage of being parasitized. *Nature* 219:690–694.

Smole, William J. 1976. *The Yanoama Indians: a cultural geography.* University of Texas Press, Austin.

Soudsky, B., and I. Pavlu. 1971. The linear pottery culture settlement patterns of Central Europe. In *Man, Settlement, and Urbanism*, edited by P. J. Ucko, R. Tringham, and G. W. Dimbleby. Duckworth, London. Pp. 317–328.

Speth John D., and Katherine A. Spielmann. 1982. Energy source, protein metabolism and hunter-gatherer subsistence strategies. *Journal of Anthropological Archaeology* 2:1–31.

Spielmann, Katherine A. 1982. *Inter-Societal Food Acquisition Among Egalitarian Societies: an ecological study of Plains/Pueblo interaction in the American Southwest.* University Microfilms, Ann Arbor, MI.

______. 1986. Interdependence among egalitarian societies. *Journal of Anthropological Archaeology* 5:279–312.

Starling, Nicholas J. 1983. Neolithic Settlement Patterns in Central Germany. *Oxford Journal of Archaeology* 2(1):1–12.

______. 1985. Colonization and succession: the earlier Neolithic of Central Europe. *Proceedings of the Prehistoric Society* 51:41–57.

Startin, W. 1978. Linear pottery culture houses: reconstruction and manpower. *Proceedings of the Prehistoric Society* 44:143–159.

Statistisch-Topographisches Bureau. 1850–1905. *Württembergische Jahrbücher für Statistik und Landeskunde.* Annual volumes 1850–1905. Lindemann Verlag, Stuttgart.

Steensberg, Axel. 1979. *Draved: an experiment in stone age agriculture, burning, sowing, and harvesting.* The National Museum of Denmark, Copenhagen.

______. 1980. *New Guinea Gardens.* Academic Press, New York.

Stevens, G. A. 1932. The number and weight of seeds produced by weeds. *American Journal of Botany* 19(9):784–794.

Tanno, Tadashi. 1976. The Mbuti net hunters in the Ituri Forest, Eastern Zaire—their hunting activities and band composition. *Koyoto University African Studies* 10:101–135.

Taute, W. 1966. Das Felsdach Lautereck, eine mesolithisch-neolithisch-bronzezeitliche Stratigraphie an der Oberen Donau. *Paleohistoria* 12: 483–504.

______. 1967 Grabungen zur mittleren Steinzeit in Höhlen und unter Felsdächern der Schwäbischen Alb 1961-1965. *Fundberichte aus Schwaben.* New Series 18/I: 14–21.

______. 1974. Neolithische Mikrolithen und andere neolithische Silexartefackte aus Süddeutschland und Österreich. *Archäologische Informationen* 3:71–77.

______. 1977. Zur Problematik von Mesolithikum und Frühneolithikum am Bodensee. In *Bodman, Dorf Kaiserpfalz Adel* Vol. 1 , edited by H. Berner. Bodenseebibliothek, Sigmaringen. Pp. 11–32.

______. 1980. *Das Mesolithikum in Süddeutschland. Teil 2: Naturwissenschaftliche Untersuchungen.* Verlag Archaeologica Venatoria, Tübingen.

Titiev, Mischa. 1944. *Old Orabi.* Papers of the Peabody Museum of American Archaeology and Ethnology No. 22. Harvard University Press, Cambridge.

Thomas, Jack W., and Dale E. Toweill (eds.). 1982. *Elk of North America: ecology and management.* United States Department of Agriculture, Forest Service. Stackpole Books, Boston.

Thompson, John N. 1982. *Interaction and Coevolution.* Wiley, New York.

Tkachuk, Russell, and V. Jean Mellish. 1977. Amino acid and proximate analysis of weed seeds. *Canadian Journal of Plant Science* 57(1):243–249.

Torke, Wolfgang. 1981. *Fischreste als Quellen der Ökologie und Ökonomie in der Steinzeit Südwest-Deutschlands.* Urgeschichtliche Materialhefte 4. Verlag Archaeologica Venatoria, Tübingen.

Trager, W. 1970. *Symbiosis.* Van Nostrand, New York.

Trigger, Bruce. 1976. *The Children of Aataentsic: a history of the Huron People.* McGill-Queens University Press, Montreal.

Tringham, Ruth. 1968. A preliminary study of the early neolithic and latest mesolithic blade industries in Southeast and Central Europe. In *Studies in Ancient Europe: essays presented to Stuart Piggott*, edited by J. M. Coles and D. D. A. Simpson. Leicester University Press, Leicester. Pp. 45–71.

_______. 1971. *Hunters, Fishers, and Farmers of Eastern Europe 6000–3000 BC.* Huchinson and Co., London.

Uerpmann, Margaret. 1976. *Zur Technologie und Typologie neolithischer Feuersteingeräte.* Tübinger Monographie 2. Verlag Archaeologica Venatoria, Tübingen.

Vandermeer, J. H. 1980. Indirect mutualism: variations on a theme by Stephen Levine. *American Naturalist* 116:441–448.

Vandermeer, J. H., and D. H. Boucher. 1978. Varieties of mutualistic interaction in population models. *Journal of Theoretical Biology* 74:549–558.

Van Dyne, G., N. Brockington, Z. Szocos, J. Duek, and C. Ribic. 1980. Large herbivore subsystems. In *Grasslands, Systems Analysis and Man.* International Biological Program 19. Cambridge University Press, London.

Vaughn, J. G. 1980. *The Structure and Utilization of Oil Seeds.* Chapman and Hall Ltd., London.

Vierich, Helga I. D. 1982. Adaptive flexibility in a multi-ethnic setting: the Basarwa of the Southern Kalahari. In *Politics and History in Band Societies*, edited by E. Leacock and R. Lee. Cambridge University Press, Cambridge. Pp. 213–222.

Wagner, G. 1986 Zur Entstehung der steinzeitlichen Fundplätze im Federseegebiet und ihre Beeinflussung durch geologische Abtragung und Landschaftsveränderung durch den Menschen. In *Archäologische Ausgrabungen in Baden-Württemberg 1985* edited by Dieter Plank. Konrad Theiss Verlag, Stuttgart. Pp. 30–33.

Wassermann, L. 1967. Hülsenfrüchte. In *Handbuch der Lebensmittelchemie* 5(1):394–417.

Waterbolk, H. T. 1968. Food production in prehistoric Europe. *Science* 162:1093–1102.

Watt, Bernice K., and Annabell Merrill. 1975. *Composition of Foods.* Agriculture Handbook No. 8. United States Department of Agriculture, Washington DC.

Weiss, Kenneth M. 1973. *Demographic Models for Anthropology.* Memoirs of the Society for American Archaeology No. 27.

WHO. 1974. *Handbook on Human Nutritional Requirements.* World Health Organization. Monograph Series 61. World Health Organization: Geneva.

Willerding, Ulrich. 1979 Paläo-ethnobotanische Untersuchungen über die Entwicklung von Pflanzengesellschaften. In *Werden und Vergehen von Pflanzengesellschaften,* edited by Ottie Wilmanns and R. Tüxen. J. Cramer Verlag, Braunschweig. Pp. 61–109.

______. 1980. Zum Ackerbau der Bandkeramiker. *Materialhefte zur Ur- and Frühgeschichte Niedersachsens* 16:421–456.

______. 1981. Ur- und frühgeschichtliche sowie mittelalterliche Unkrautfunde in Mitteleuropa. *Zeitschrift für Pflanzenkrankheiten und Pflanzenschutz* 9:65–74.

______. 1983. Zum ältesten Ackerbau in Niedersachsen. In *Frühe Bauernkulturen in Niedersachsen.* Archäologische Mitteilungen aus Nordwestdeutschland, Beiheft I, edited by Günter Wagner. Pp. 179–219.

Willfort, Richard. 1982. *Gesundheit durch Heilkräuter.* Twenty-second edition. Trauner Verlag, Linz.

Williams, B. J.1974. *A Model of Band Society.* American Antiquity Memoir No. 29.

Wills, W. H. 1985. *Early Agriculture in the Mogollon Highlands of New Mexico.* University Microfilms, Ann Arbor, MI.

Winterhalder, B., and E. A. Smith. 1981. *Hunter-gatherer Foraging Strategies: ethnographic and archaeological analyses.* University of Chicago Press, Chicago.

Wobst, H. Martin. 1974. Boundary conditions for Paleolithic social systems: a simulation approach. *American Antiquity* 39(2):147–179.

Wolin, Carole L., and Lawrence R. Lawlor. 1984. Models of facultative mutualism: density effects. *American Naturalist* 124(6):843–862.

Wright, Gary A. 1967. Some aspects of early and mid-seventeenth century exchange networks in the western Great Lakes. *Michigan Archaeologist* 13(4):181–196.

Yarnell, Richard A. 1964. *Aboriginal relationships between culture and plant life in the Upper Great Lakes region.* Anthropological Papers 24. Museum of Anthropology, Ann Arbor, MI.

Works Cited

Zeist, W. van. 1970. Prehistoric and early historical food plants in the Netherlands. *Paleohistoria* 24:41–175.

Zimmerman, Martin. 1950. *Schilpfs praktisches Handbuch der Landwirtschaft*. Parey Verlag, Berlin.

Index

Index

Index

Index

Index

Index